LEARNING & MEMORY

Editor
John H. Byrne (Houston)

Managing Editor
Judy Cuddihy (Cold Spring Harbor)

Editorial Board

Per Anderson (Oslo)	Peter Holland (Durham)
Philippe Ascher (Paris)	Eric Kandel (New York)
Jocelyne Bachevalier (Houston)	Lawrence C. Katz (Durham)
Alan D. Baddeley (Cambridge)	Mary B. Kennedy (Pasadena)
Carol A. Barnes (Tucson)	Joseph Le Doux (New York)
Timothy Bliss (London)	Stephen G. Lisberger (San Francisco)
Thomas J. Carew (New Haven)	Nicholas J. Mackintosh (Cambridge)
Graham Collingridge (Birmingham)	Daniel Madison (Stanford)
John Connor (Albuquerque)	Roberto Malinow (Cold Spring Harbor)
Thomas Curran (Memphis)	Randolf Menzel (Berlin)
Antonio Damasio (Iowa City)	Mortimer Mishkin (Bethesda)
Michael Davis (New Haven)	Richard Morris (Edinburgh)
Ronald Davis (Houston)	Dennis D.M. O'Leary (La Jolla)
Pietro De Camili (New Haven)	Marcus Raichle (St. Louis)
Yadin Dudai (Rehovot)	Christine Sahley (West Lafayette)
Howard Eichenbaum (Stony Brook)	Daniel Schacter (Cambridge)
Yves Frégnac (Gif sur Yvette)	James Schwartz (New York)
Alan Gelperin (Murray Hill)	Carla Shatz (Berkeley)
Alison Goate (St. Louis)	Wolf Singer (Frankfurt)
Patricia Goldman-Rakic (New Haven)	Larry Squire (San Diego)
Michael E. Greenberg (Boston)	Charles Stevens (La Jolla)
Stephen Heinemann (La Jolla)	Richard Thompson (Los Angeles)
Martin Heisenberg (Wurzburg)	Richard Tsien (Stanford)
Susan Hockfield (New Haven)	Tim Tully (Cold Spring Harbor)

Editorial Offices
Cold Spring Harbor Laboratory Press
1 Bungtown Road
Cold Spring Harbor, New York 11724
Phone (516) 367-8492
Fax (516) 367-8532

Editorial/Production
Nadine Dumser, Technical Editor
Kristin Kraus, Production Editor
Cindy Grimm, Production Assistant
Doris Lawrence, Editorial Secretary

Learning & Memory (ISSN 1072-0502) is published bimonthly for $210 (U.S. institutional; $225 rest of world; $245 R.O.W. with airlift), $95 (individual making personal payment; $110 R.O.W. surface; $130 with airlift) by Cold Spring Harbor Laboratory Press, 1 Bungtown Road, Cold Spring Harbor, New York 11724. Periodicals postage pending is paid at Cold Spring Harbor and additional mailing offices. POSTMASTER: Send address changes to Cold Spring Harbor Laboratory Press, 10 Skyline Drive, Plainview, New York 11803-2500. **Subscriptions:** Barbara Terry, Subscription Manager. Personal: U.S. $95, R.O.W. $110 surface mail, $130 with airlift delivery. Institutional: U.S. $210; R.O.W. $225 surface mail, $245 with airlift delivery. Orders may be sent to Cold Spring Harbor Laboratory Press, Fulfillment Department, 10 Skyline Drive, Plainview, New York 11803-2500. Telephone: Continental U.S. and Canada 1-800-843-4388; all other locations 516-349-1930. FAX: 516-349-1946. Personal subscriptions must be prepaid by personal check, credit card, or money order. Claims for missing issues must be received within 4 months of issue date.
Advertising: Marcie Ebenstein, Advertising Manager, Cold Spring Harbor Laboratory Press, 1 Bungtown Road, Cold Spring Harbor, New York 11724-2203. Phone: 516-367-8351. FAX: 516-367-8532.
Copyright information: Authorization to photocopy items for internal or personal use of specific clients is granted by Cold Spring Harbor Laboratory Press for Libraries and other users registered with the Copyright Clearance Center (CCC) Transactional Reporting Service, provided that the base fee of $5.00 per copy is paid directly to CCC, 21 Congress Street, Salem, Massachusetts 01970 (1072-0502/97 + $5.00). This consent does not extend to other kinds of copying, such as copying for general distribution for advertising or promotional purposes, for creating new collective works, or for resale.

Copyright © 1997 by Cold Spring Harbor Laboratory Press

The Latest Volume From The Most Prestigious Book Series in Biology

Function & Dysfunction in the Nervous System

Cold Spring Harbor Symposia on Quantitative Biology, Volume LXI

Much is being learned about how the brain works by studies of diseases that affect it. This ambitious volume portrays some of the most exciting aspects of neuroscience today - language development, visual awareness, neuronal plasticity, sensory perception, memory formation - by presenting studies on their normal mechanisms alongside illuminating investigations of abnormalities caused by degenerative disease, addiction, developmental errors, and other maladies. In this sixty-first volume of the most prestigious book series in experimental biology, over eighty of the world's most distinguished investigators provide a perspective that resonates beyond the lab bench into the clinics of the not-so-distant future.

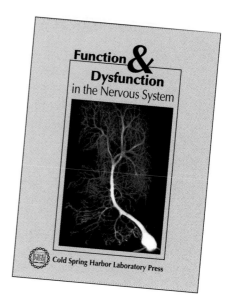

CONTENTS (and principal authors)

Language and Visual Processing
M.M. Merzenich, M.F. Raichle, R.A. Andersen, D.H. Hubel, J.A. Movshon, C. Koch

Handedness
A.J.S. Klar

Perceptual Plasticity
S. Nakanishi, T. Bonhoeffer, L.C. Katz, H.T. Cline, C.D. Gilbert, V.S. Ramachandran

Sensory Perception
R. Axel, L.B. Buck, C.I. Bargmann, R.R. Reed, R.F. Margolskee

Memory
L.R. Squire, H. Eichenbaum, T. Tully, E.R. Kandel, S. Tonegawa, A.J. Silva, J.E. LeDoux, L. Cahill

Rhythms
M. Rosbash, M.W. Young, P. Sassone-Corsi, J.S. Takahashi

Channels and Receptors
A.J. Hudspeth, D.P. Corey, J.L. Noebels, J.O. McNamara, P.R. Schofield, J.-P. Changeux, Z.W. Hall, M.V.L. Bennett

Neuronal Survival and Signaling
D.W. Choi, W.C. Mobley, Y.-A. Barde, G. Enikolopov, G.D. Fischbach, G.D. Yancopoulos, H. Westphal, M.A. Marchionni

Prions
S.B. Prusiner, C. Weissmann, J.W. Ironside, S.J. DeArmond, R.B. Wickner

Alzheimer's Disease
A. Goate, P.H. St. George-Hyslop, M. Goedert, C.L. Masters, D.J. Selkoe, A.D. Roses

Neurodegenerative Diseases
C.T. Caskey, J.F. Gusella, A.L. Joyner, J.-L. Mandel, H.T. Orr, J.R. Lupski, K.H. Fischbeck, S.T. Warren, G. Dreyfuss, T. Siddique, D.W. Cleveland, D.L. Price, S. Strickland, R.H. Edwards

Dementia, Addiction, and Psychiatric Diseases
R.W. Price, T.N. Ferraro, J.R. DePaulo, Jr., G.M. Goodwin, M. Karayiorgou, P. McGuffin, R.E. Straub, A.E. Pulver, D.B. Wildenauer, Z.W. Hall

1997, 875 pp., illus., color plates, appendices
Cloth $230   ISBN 0-87969-071-2
Paper $99   ISBN 0-87969-072-0

How to Order!

Call Toll-Free:	1-800-843-4388 (Continental U.S. and Canada)
or Call:	516-349-1930 (All other locations)
FAX:	516-349-1946
E-mail:	cshpress@cshl.org
WWW Site:	http://www.cshl.org/
Write:	CSHL Press, 10 Skyline Drive, Plainview, NY 11803-2500

From  **Cold Spring Harbor Laboratory Press**

Volume 4
Number 1

May/June 1997
Pages 1-178

Review

Prediction and Preparation, Fundamental Functions of
the Cerebellum 1
Eric Courchesne and Greg Allen

Research papers

Impaired Capacity of Cerebellar Patients to Perceive and Learn
Two-Dimensional Shapes Based on Kinesthetic Cues 36
Yury Shimansky, Marian Saling, David A. Wunderlich, Vlastislav Bracha,
George E. Stelmach, and James R. Bloedel

Lateral Cerebellar Hemispheres Actively Support Sensory
Acquisition and Discrimination Rather Than Motor Control 49
Lawrence M. Parsons, James M. Bower, Jia-Hong Gao, Jinhu Xiong, Jinqi Li, and
Peter T. Fox

Cerebellar Guidance of Premotor Network Development and
Sensorimotor Learning 63
Sherwin E. Hua and James C. Houk

Role of Cerebellum in Adaptive Modification of Reflex Blinks 77
John J. Pellegrini and Craig Evinger

Single-Unit Evidence for Eye-Blink Conditioning in Cerebellar
Cortex is Altered, but Not Eliminated, by Interpositus
Nucleus Lesions 88
Donald B. Katz and Joseph S. Steinmetz

Effect of Varying the Intensity and Train Frequency of Forelimb
and Cerebellar Mossy Fiber Conditioned Stimuli on the Latency
of Conditioned Eye-Blink Responses in Decerebrate Ferrets 105
Pär Svensson, Magnus Ivarsson, and Germund Hesslow

Conditioned Response Timing and Integration in
the Cerebellum 116
John W. Moore and June-Seek Choi

A Model of Pavlovian Eyelid Conditioning Based on the
Synaptic Organization of the Cerebellum 130
Michael D. Mauk and Nelson H. Donegan

Local Dendritic Ca^{2+} Signaling Induces Cerebellar
Long-Term Depression 159
Jens Eilers, Hajime Takechi, Elizabeth A. Finch, George J. Augustine, and
Arthur Konnerth

Absence of Cerebellar Long-Term Depression in Mice Lacking
Neuronal Nitric Oxide Synthase 169
Varda Lev-Ram, Zuryash Nebyelul, Mark H. Ellisman, Paul L. Huang, and
Roger Y. Tsien

Cover Dissociation of cerebellar attention (yellow and blue) and motor (green and red) activation (yellow and green = overlap in activation of 3 or more subjects; blue and red = overlap of any 2 subjects). Three-dimensional volume rendering of the cerebellum and brain stem demonstrates that during an attention task, the most common site of activation was in the left superior posterior cerebellum, while during a motor task, the most common site was in the right anterior cerebellum. (For details, see Courchesne and Allen, p. 1; image rendered using VoxelView 2.5.)

The following articles appeared last month in the first special issue devoted to learning and the cerebellum, *Learning & Memory*, vol. 3, number 6, March/April 1997

Review

The Cerebellum, LTD, and Memory: Alternative Views
Rodolfo Llinás, Eric J. Lang, and John P. Welsh

Research papers

Preserved Performance by Cerebellar Patients on Tests of Word Generation, Discrimination Learning, and Attention
Laura L. Helmuth, Richard B. Ivry, and Naomi Shimizu

A Neural Model of Cerebellar Learning for Arm Movement Control: Cortico-Spino-Cerebellar Dynamics
Jose L. Contreras-Vidal, Stephen Grossberg, and Daniel Bullock

Multiple Subclasses of Purkinje Cells in the Primate Floccular Complex Provide Similar Signals to Guide Learning in the Vestibulo-ocular Reflex
Jennifer L. Raymond and Stephen G. Lisberger

The Effects of Reversible Inactivation of the Red Nucleus on Learning-Related and Auditory-Evoked Unit Activity in the Pontine Nuclei of Classically Conditioned Rabbits
M. Claire Cartford, Elizabeth B. Gohl, Maria Singson, and David G. Lavond

The Learning-Related Activity That Develops in the Pontine Nuclei During Classical Eye-Blink Conditioning Is Dependent on the Interpositus Nucleus
Robert E. Clark, Elizabeth B. Gohl, and David G. Lavond

Reversible Inactivation of the Cerebellar Interpositus Nucleus Completely Prevents Acquisition of the Classically Conditioned Eye-Blink Response
David J. Krupa and Richard F. Thompson

Acquisition of a New-Latency Conditioned Nictitating Membrane Response—Major, But Not Complete, Dependence on Ipsilateral Cerebellum
Christopher H. Yeo, Dominic H. Lobo, and Alison Baum

Persistent Phosphorylation Parallels Long-Term Desensitization of Cerebellar Purkinje Cell AMPA-Type Glutamate Receptors
Kazutoshi Nakazawa, Sumiko Mikawa, and Masao Ito

SPRING TITLES from Cold Spring Harbor Laboratory Press

Fly Pushing
The Theory and Practice of *Drosophila* Genetics

By Ralph J. Greenspan, *New York University*

Entertaining and lucid, Greenspan guides the reader carefully through the practicalities of making crosses, isolating variants, mapping genes, constructing specific genotypes, and analyzing mutations. No previous knowledge of fly genetics is assumed. The techniques used are illustrated, and practice problems and solutions are included to assist the reader.

1997, 155 pp., illus., appendices index
Cloth $35   ISBN 0-87969-492-0

Epigenetic Mechanisms of Gene Regulation
Monograph 32

Edited by Vincenzo E.A. Russo, *Max-Planck-Institut für Molekulare Genetik*; Robert A. Martienssen, *Cold Spring Harbor Laboratory*; Arthur D. Riggs, *Beckman Research Institute of the City of Hope*

This monograph, edited by three well-known biologists from different specialties, is the first to review and synthesise what is known about these effects across all species, particularly from a molecular perspective, and will be of interest to everyone in the fields of molecular biology and genetics.

1996, 692 pp., illus., color plates, glossary, index
Cloth $125   ISBN 0-87969-490-4

Oxidative Stress and the Molecular Biology of Antioxidant Defenses
Monograph 34

Edited by John G. Scandalios, *North Carolina State University*

Written and edited by leaders in this growing field, this volume is an essential work of reference for specialists and investigators with wider interests in cell biology, aging and cancer biology.

1997, 904 pp., illus., color plates, index
Cloth $150   ISBN 0-87969-502-1

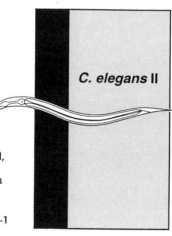

C. elegans II
Monograph 33

Edited by Donald L. Riddle, *University of Missouri, Columbia*; Thomas Blumenthal, *Indiana University*; Barbara J. Meyer, *University of California, Berkeley*; and James R. Priess, *Howard Hughes Medical Institute, Fred Hutchinson Cancer Research Center, Seattle*

This volume is a *must* for any investigator doing worm studies but it has been written and rigorously edited for a wider community of investigators in cell and molecular biology who should know how new knowledge of *C. elegans* relates to their own specialty.

1997, 1222 pp., illus., color plates, index
Cloth $175   ISBN 0-87969-488-2

Genetic Instability in Cancer
Cancer Surveys, Volume 28

Edited by Thomas Lindahl, *Imperial Cancer Research Fund*

This series, edited by J. Tooze (as of Volume 26), provides a comprehensive survey of the present state and future developments in well-defined areas in oncology. Each issue deals with a specific topic and has guest editors with an expert knowledge of the subject.

1996, 353 pp., illus., color plates index
Cloth $90   ISBN 0-87969-485-8

Correcting the Blueprint of Life
An Historical Account of the Discovery of DNA Repair Mechanisms

By Errol C. Friedberg, *The University of Texas Southwestern Medical Center, Dallas*

In this brief, readable, and revealing book, one of the pioneers of the now rapidly evolving field of DNA repair traces the history of the discovery of the more important mechanisms by which cells respond to DNA damage. Errol Friedberg has written an enjoyable and informative introduction to the study of DNA mutagenesis and repair that will interest students at an advanced undergraduate or graduate student level as well as investigators in fields as diverse as oncogenesis, cell cycle regulation, transcription and DNA replication.

Due May 1997, 230 pp. (approx.), illus., index
Cloth $49   ISBN 0-87969-507-2

To order or request additional information:
Call: 1-800-843-4388 (Continental US and Canada) 516-349-1930 (All other locations)
FAX: 516-349-1946
E-mail: cshpress@cshl.org or World Wide Web Site http://www.cshl.org
Write: CSHL Press, 10 Skyline Drive, Plainview, NY 11803-2500

REVIEW

Eric Courchesne[1,2,4] and
Greg Allen[2,3]

[1]Department of Neurosciences
School of Medicine
University of California, San Diego
La Jolla, California 92093
[2]Research on the Neuroscience of Autism
Children's Hospital Research Center
La Jolla, California 92037
[3]San Diego State University/
University of California, San Diego
Joint Doctoral Program in Clinical
Psychology
San Diego, California 92120

Prediction and Preparation, Fundamental Functions of the Cerebellum

Introduction

For well over a century, scientists have continually questioned the role of the cerebellum in the central nervous system (CNS) (for review, see Schmahmann 1997). Traditionally viewed as a structure dedicated to motor control, indications to the contrary have never ceased to emerge in a variety of experimental and clinical contexts. In recent years, the burgeoning use of functional neuroimaging methodology—namely, positron emission tomography (PET) and, more recently, functional magnetic resonance imaging (fMRI)—has led to a substantial increase in findings implicating the cerebellum in various nonmotor as well as motor operations. It now appears quite clear that the cerebellum is involved in a wide range of motor, sensory, and cognitive functions. However, as for the actual role the cerebellum serves, the question remains.

Certain new concepts of the cerebellum have attempted to account for its apparent functional diversity. For instance, one view suggests that the cerebellum guides movements that reposition sensory receptors to optimize the quality of sensory information that the nervous system acquires (Bower and Kassel 1990). Although improving on older notions, this idea still views the ultimate function of the cerebellum as one of motor control; rather than freeing us from the binds of tradition, it is simply a variation on the motor theme. Another view suggests that the cerebellum is a system that tracks the trajectories of sensory information moving through the environment. Such tracking is thought to facilitate the efficiency of movement relative to or in response to moving stimuli (Paulin 1993). Although it is somewhat broader in its ability to account for the functional diversity of the cerebellum, this view is inconsistent with findings indicating that the cerebellum is involved whether the information it is acquiring is moving or not. Thus, although they are improvements on tradition, most recent attempts to reconceptualize the role of the cerebellum are too limited in scope to account for the breadth of existing data.

There is little reason to believe that the actual function of the cerebellum would be as limited as most concepts of cerebellar function

[4]Corresponding author.

might suggest. In terms of the number of neurons it contains, the cerebellum is the largest structure in the human brain (Williams and Herrup 1988). This tremendous number of neurons, coupled with the high input-to-output axon ratio (cerebellar afferents to efferents are 40:1; Carpenter 1991) suggests that its function must be massively integrative. It is also one of the most widely connected structures, having physiological connections with all major divisions of the CNS (Moruzzi and Magoun 1949; Snider 1950, 1967; Bava et al. 1966; Sasaki et al. 1972, 1979; Kitano et al. 1976; Watson 1978; Itoh and Mizuno 1979; Newman and Reza 1979; Saint-Cyr and Woodward 1980a,b; Vilensky and Van Hoesen 1981; Crispino and Bullock 1984; Ito 1984; Haines and Dietrichs 1987; King 1987; Nieuwenhuys et al. 1988; Schmahmann and Pandya 1989; Ghez 1991; Ikai et al. 1992; Llinás and Sotelo 1992; Thielert and Thier 1993; Glickstein et al. 1994; Lynch et al. 1994; Middleton and Strick 1994, Schmahmann 1996).

Moreover, experimental data and, in particular, the results of recent fMR imaging and PET studies, indicate that the cerebellum might be involved in a wide range of functions, including attention, associative learning, practice-related learning, procedural learning, declarative memory, working memory, semantic association, conditioned anxiety, mental exploration, and complex reasoning and problem solving as well as sensory, motor and motor skill acquisition (see Table 1). A general theory must be able to encompass and explain the functional role played by the cerebellum in these diverse motor and nonmotor domains.

A More Encompassing Theory

New concepts of cerebellar function are obliged to account for new facts and findings. Such a concept, which has been proposed previously (Courchesne et al. 1994; Akshoomoff et al. 1997; Allen et al. 1997; Courchesne 1997), suggests that the fundamental purpose of the cerebellum is *to predict internal conditions needed* for a particular mental or motor operation and *to set those conditions* in preparation for the operation at hand. This cerebellar preparatory function is neither a sensory nor a motor activity but, rather, a general one that prepares whichever neural systems (e.g., sensory, motor, autonomic, memory, attention, affective, speech, language) may be needed in upcoming moments. In other words, the cerebellum prepares internal conditions (e.g., by repositioning sensory receptors; by altering cerebral blood flow levels; by enhancing neural signal to noise; by enhancing neural responsiveness in hippocampus, thalamus, and superior colliculus; by modulating motor control sytems) for imminent information acquisition, analysis, or action. Its preparatory actions thereby facilitate and improve sensory processing and mental and motor performance in response to subsequent sensory events.

To perform this function, the cerebellum must learn the predictive relationships among temporally ordered multidimensional sequences of exogenously derived (e.g., sensory events) and endogenously derived (e.g., signals from prefrontal and posterior parietal cortex, hippocampus, hypothalamus, etc.) neural activities, including those derived from the consequences of its own output—preparatory signaling. Whenever an analogous sequence begins to unfold in real time, the cerebellum predicts—based on such prior learning—what is about to happen and triggers preparatory actions (from neural to cerebrovascular; from motor to mental) that alter neural response thresholds and readiness in systems

Table 1: *PET and fMRI studies show that the cerebellum is active*

During	References
Attention	Allen et al. (1997); T.H. Le and X. Hu (unpubl.)
Sensory discrimination	Gao et al. (1996)
Semantic association	Petersen et al. (1989); Martin et al. (1995)
Working memory	Awh et al. (1995); Courtney et al. (1996); Fiez et al. (1996); Klingberg et al. (1996); J.E. Desmond, J.D.E. Gabrieli, B.L.Ginier, J.B. Demb, A.D. Wagner, D.R Enzman, and G.H. Glover (unpubl.)
Associative learning	Molchan et al. (1994); Logan and Grafton (1995); Blaxton et al. (1996)
Practice-related learning	Raichle et al. (1994)
Procedural learning	Grafton, et al. (1994); Jenkins et al. (1994); Flament et al. (1996); S.L. Rao, D.L. Harrington, K.Y. Haaland, J.A. Bobholz, J.R. Binder, T.A. Hameke, J.A. Frost, B.M. Myklebust, R.D. Jacobson, P.A. Bandettini, and J.S. Hyde (unpubl.)
Motor skill acquisition	Friston et al. (1992); Seitz et al. (1994)
Problem solving	Kim et al. (1994)
Concept formation	Berman et al. (1995); Nagahama et al. (1996)
Conditioned anxiety	A.L. Malizia, S.J. Wilson, J.-B. Poline, D.J. Nutt, and P.M. Grasby (unpubl.)
Mental exploration	Mellet et al. (1995)
Spatial memory	Moscovitch et al. (1995)
Object memory	Moscovitch et al. (1995)
Verbal memory	Grasby et al. (1993); Andreasen et al. (1995)
Episodic memory	Andreasen et al. (1995)
Semantic memory	Andreasen et al. (1995)
Speech	E. Artiges, M.J. Giraud, B. Mazoyer, C. Trichard, L. Mallet, H. de la Caffiniere, M. Verdys, A.M. Syrota, and J.L. Martinot (unpubl.)
Motor imagery	Decety et al. (1994); Parsons et al. (1995)
Motor preparation	Deiber et al. (1996)
Motor control	Fox et al. (1985); Ellerman et al. (1994); Allen et al. (1997)

expected to be needed in upcoming moments. Complete knowledge of upcoming events is not necessary to trigger specific preparatory actions. What is important is the probabilistic relevance of a particular sequence or stimulus as a predictor of an upcoming event (Coenen and Sejnowski 1996; their computational model of the cerebellum as a "predictive machine" is discussed below). Simple exposure to aspects of a sequence of activities (even a single exogenous or endogenous event) that are predictive of events that will soon arrive may be sufficient to trigger preparatory responding by the cerebellum. The consequences of this responding are fed back to the cerebellum and modify these temporally ordered multidimensional predictive sequences and the cerebellar preparatory responses that they engender. In this way, the predictions and preparatory signals of the cerebellum are adjusted to meet changing internal and external conditions. The end result is a dynamically modifiable prediction and preparatory response representation.

Cerebellar learning, then, is not strictly speaking motor learning, declarative learning, or procedural learning. Rather, it is preparatory learning, the purpose of which is to predict and prepare *internal conditions* needed to facilitate the efficient and timely responses of a wide variety of motor and nonmotor systems. Such learned associative responses provide automatic, moment-to-moment signaling. To be maximally efficient and timely, the cerebellum must be able to trigger, via signals to different neural systems (autonomic, sensory, attention, etc.), preparation response enhancement periods of variable duration from tens of milliseconds to minutes, and it must also be able to effect specific local neural changes as well as nonspecific global system-wide changes. An exact knowledge of upcoming sensory and neural sequences can improve the precision of preparatory responses, and experimenters can deliberately or unwittingly supply this sort of information in test situations. However, in the natural world, it is unusual to have such a complete, conscious foreknowledge of upcoming events. To be maximally adaptive to real world variability, then, the cerebellum must be a "pattern extractor," capable of getting the "gist" of what has been happening, what is likely to happen next, and what internal conditions are needed to prepare for a particular predicted up-coming operation, be it acquisition, analysis, or action. Interestingly, if one were to name a single characteristic of the cognitive style of patients with autism, a developmental disorder involving prenatal loss of Purkinje neurons (Courchesne 1997), it would be that they are literal-minded and unable to get the gist of things [e.g., see the very first description of an autistic child (Kanner 1943)].

Without the specific type of automatic yet modifiable, moment-to-moment preparatory aid provided by the cerebellum, other systems—motor, sensory, attention, language, and so forth—could continue to perform their prescribed specific functions but would do so suboptimally in situations in which prediction and preparation facilitate performance. For instance, observations and experiments show that in situations demanding movement, perception, attention, or association learning, cerebellar damage does not eliminate function, but it does increase suboptimal variability in response thresholds, times, and amplitudes, and it does increase conscious effort when performing motor or mental tasks. Gordon Holmes (1939) quotes a cerebellar-lesioned patient as saying, "The movements of my left (unaffected) arm are done subconsciously, but I have to think out each movement of the right (affected) arm. I come to a dead stop in turning and have to think before I start again." In short, in this patient, automatic, continuous, and unconscious preparatory facilitation was eliminated, and he had to operate, metaphorically speaking, "in the present tense." The cerebellar pathology did not eliminate the patient's voluntary motor action but, instead, made motor action slow, inaccurate, and effortful; the patient had to consciously "think" through each step in preparation for action and during the execution of each action. In a parallel fashion, evidence shows that cerebellar pathology does not eliminate voluntary shifts of attention but, instead, makes such shifts slow and inaccurate (Akshoomoff and Courchesne 1992, 1994; Courchesne et al. 1994). Similarly, cerebellar pathology apparently does not prevent relatively good perceptual judgments of time intervals between any two stimuli but, instead, makes such judgments more variable. For instance, whereas normal subjects and patients with cerebellar damage differ from each other by as little as 1% in correctly tapping out a 550-msec time interval between two tones, the patients show a significant increase relative to normals in the standard deviations associated with such performance or perceptual judgments (±45.7 vs. ±26.1 msec) (Ivry and Keele 1989). Likewise, cerebellar lesions do not necessarily eliminate the classically conditioned nictitating membrane response, but they do produce variability in response onset latency and amplitude (e.g., Fig. 16.13 from Welsh and Harvey 1992). We suggest that in all of the above cases, damage to the cerebellum disrupted the normal preparation of other neural systems, thereby impairing performance.

It should be noted that this is not the first concept of cerebellar function to include the element of prediction. Certain models that have ascribed a predictive role to the cerebellum have done so as a means of explaining the role that the cerebellum plays in motor control (e.g., Darlot 1993; Miall et al. 1993). In their paper describing the cerebellum as a Smith predictor, Miall et al. (1993) suggest "that the cerebellum may be involved in more complex predictions, linking it to more cognitive processes." Also, in his model of cerebellar function described above, Paulin (1993) proposes that the cerebellum improves the efficiency of motor control through the process of "state estimation," which might involve prediction. Finally, Coenen and Sejnowski (1996) have begun to consider the physiological foundation for the formation of predictions in the cerebellum. In our theory, and in this paper, we consider how predictions generated in the cerebellum might be implemented in order to prepare diverse neural systems, both motor and nonmotor.

To this date, there have been no experiments that directly test the specific elements of the proposed theory of cerebellar function. However, two contexts in which such a function is predicted to be most evident are studies of attention and studies of learning, two areas in which cerebellar involvement has been investigated extensively.

Cerebellar Involvement in Attention and Preparation

A novel approach to elucidating the role of the cerebellum in human neural functioning is through examination of its role in the coordination of attention and other preparatory functions. Because attention, sensory response modulation, and related preparatory functions can operate independently of motor control, studies of attention offer a new way of looking at the nonmotor functions of the cerebellum. This novel perspective may increase the chances of discovering the fundamental properties of cerebellar function that are common across and underlie its influence on different neurobehavioral domains.

It is common sense that paying attention "after the fact" is too late (you have already run the red light or missed what the teacher said), and that conversely the benefits of attention accrue when it is properly directed and applied before something important occurs. That way, one is properly prepared to sense, analyze, and act. Of course, to direct attention properly, the brain must guess from past experience, learning, and knowledge of current events what is most likely to happen next.

Attention, then, is an act of preparation, and involves the selective modulation of neural responsiveness in many systems in advance of anticipated sensory information (see Animal Studies, below). So, when sensory information is anticipated that may have signal value to the task at hand or may have imminent, inherent biological significance, attention is normally redirected toward the predicted source of that information to prepare for acquisition, analysis, or action. Thus, attention is a "preprocessing" or "advanced preparation" mechanism that facilitates and improves sensory processing and mental and motor performance in response to subsequent sensory events.

Therefore, if the theorized cerebellar function is indeed a general one that prepares whichever neural systems may be necessary in upcoming moments, then it should be true that the cerebellum plays a role in attention and other preparatory operations independent of motor involvement. In fact, a role for the cerebellum in attention operations was demonstrated by what was arguably one of the first attempts to directly test the role of the cerebellum in a specific cognitive operation, as discussed next (Courchesne et al. 1990, 1994; Akshoomoff and Courchesne 1992, 1994). That series of studies paved the way for functional neuroimaging studies of attention in the normal human cerebellum and also prompted a reexamination and reinterpretation of past studies of the cerebellum in animals and recent hypotheses (Bower and Kassel 1990; Gao et al. 1996) (see below).

EARLY STUDIES OF SHIFTING ATTENTION AND THE CEREBELLUM

Based on its privileged physioanatomical position allowing it to affect known attention systems, it was hypothesized over a decade ago that the cerebellum contributes to attention operations in a manner analogous to its role in motor control (Courchesne 1987 and unpubl.). Thus, it was predicted that the cerebellum allows attention to be shifted rapidly, accurately, smoothly, and effortlessly.

To investigate this new hypothesis, patients with acquired focal neocerebellar lesions were tested in an original paradigm (Fig. 1, top) in which cues presented at unpredictable time intervals directed patients to initiate shifts of their focus of attention between visual and auditory sources of information (Akshoomoff and Courchesne 1992; Courchesne et al. 1994) or between color and form information (Akshoomoff and Courchesne 1994). Also tested were patients with autism, a disorder involving Purkinje neuron loss (Courchesne 1997). Performance on this task was compared to performance on a task that was identical apart from the fact that it did not require subjects to shift their focus of attention. Rather, attention was sustained on a single source of information (e.g., visual) throughout the task. Several lines of neurobehavioral and neurophysiological evidence demonstrated an impaired ability to rapidly and accurately shift the mental focus of attention in neocerebellar patients and autistic patients. Evidence also showed that this impairment was not attributable to motor control deficits.

NEUROBEHAVIORAL EVIDENCE

Within 2.5 sec or less following a cue to shift attention, patients with neocerebellar lesions and autistic patients were significantly worse than normal subjects and patients with focal cerebral lesions in correctly detecting target information in the new focus (Courchesne et al. 1990;

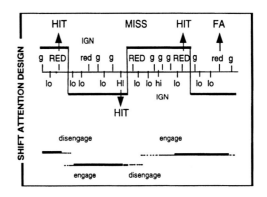

Figure 1: (*Top*) Schematic of a visual-color/auditory-tone shift attention task. Visual stimuli were red and green flashes; auditory stimuli were 2 kHz (hi) and 1 kHz (lo) tone pips. (HIT) Correctly detected target; (MISS) failure to respond to a target; (FA) an erroneous response to a rare stimulus that was in a modality to be ignored; (ign) a rare stimulus that was correctly ignored. In the example of the shift experiment, the subject pressed a button (arrow) to the first rare visual target stimulus. This served as a cue to shift attention to the auditory stimuli, ignore (IGN) the visual stimuli, and respond to the next auditory target, which in turn served as a cue to shift attention back to the visual stimuli. The subject's attentional field is indicated by the heavy line. (Adapted from Akshoomoff and Courchesne 1992). (*Bottom*) Time-related shift attention deficits in 6 patients with acquired neocerebellar lesions (▲) and 13 patients with autism (■), a disorder involving Purkinje neuron loss in early development; performance compared to 25 normal controls (●). Larger negative difference values indicate a larger decrement in performance on the shift task (see *top*) relative to a baseline control sustained attention task. This difference in the median percent hits is graphed as a function of elapsed time since the immediately preceding target. Natural log time scale in seconds. (Adapted from Courchesne et al. 1994).

Akshoomoff and Courchesne 1992; Courchesne et al. 1994) (Fig. 1). Akshoomoff and Courchesne (1994) demonstrated further that in patients with neocerebellar lesions, this deficit is present even when shifting occurs within a single visual spatial location between color and form stimuli. All of the reaction times for the neocerebellar and autistic patients were well within a 200- to 1400-msec time window allotted for responses [their median reaction times (RTs) being ~500–600 msec], eliminating the possibility that motor responses were so slow that they were not counted. Also, as long as they were not required to shift attention, that is, while attention was sustained on a single modality, these patients were not significantly impaired when responding to two target stimuli occurring rapidly in succession. That is, as long as the neocerebellar-lesioned patients did not have to change preparatory states, their attention and motor performance were not impaired.

EVENT-RELATED BRAIN RESPONSE EVIDENCE

To verify that the neocerebellar and autistic patients had not mentally shifted their attention when they missed targets, we recorded the P3b event-related potential (ERP) to all hits and misses. The P3b is a sign of covert attention independent of overt motor action (Courchesne et al. 1977), and it is absent when a target stimulus is ignored or missed (Squires et al. 1973; Ciesielski et al. 1990). Like normal subjects, the neocerebellar and autistic patients exhibited a P3b response to correctly detected targets but not to missed stimuli that occurred 2.5 sec or less following a cue. These findings suggest that when these patients missed targets that rapidly followed a cue, they were not covertly attending and thus had not fully shifted their attention to the new focus. As with the preceding behavioral observations, this neurophysiological observation also reflects a mental rather than motor output error.

To further verify the behavioral evidence showing that neocerebellar patients needed more time to completely reestablish a selective focus of attention following a cue to shift, we analyzed the P3b response to all hits as a function of time since the last target was correctly detected. Figure 2 shows the results from the task requiring shifts between color and form stimuli. In normal subjects, within the shortest time interval (as well as long ones) following a cue to shift attention to the other perceptual domain, targets in that new focus elicited a large P3b amplitude, whereas correctly ignored rare stimuli in the old focus elicited little or no P3b. This large P3b response difference between attended and ignored stimuli is nonmotor evidence that in the shifting attention task, normal subjects were able to very rapidly and selectively turn off active attention to one perceptual domain (e.g., color) and turn it on to the other (e.g., form). However, this was not the case with the neocerebellar patients (Fig. 2). Within the shortest time interval (<2.5 sec) following a cue to shift attention, targets to correctly detected and correctly rejected stimuli elicited similar amplitude P3b responses. However, given more time (2.5–30.0 sec), attentional selectivity did emerge as demonstrated by the clear amplitude difference between P3b responses to correctly detected targets and those to correctly ignored rare stimuli. This is motor-free neurophysiological evidence that neocerebellar damage impairs the ability to rapidly and selectively enhance or reduce attention.

Finally, in normal subjects the cue to execute a mental shift of attention elicits a shift attention difference, ERP, the Sd potential (Courchesne et al. 1995). Its regional distribution may vary according to whether the eliciting cue signals the need to shift attention to auditory or visual information (E. Courchesne, N. Akshoomoff, J. Townsend, and O. Saitoh, unpubl.), and so the Sd may reflect operations involved in setting up a

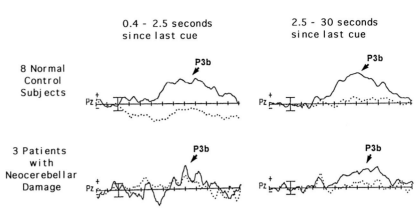

Figure 2: ERP recordings during a visual-color/visual-form shift attention task analogous to the visual-color/auditory-tone shift attention task diagrammed in Fig. 1 (*top*). Shown are time-related attention effects for three patients with neocerebellar damage (mean age, 9.7 years) and eight normal control subjects (mean age, 9.6 years). Note the minimal difference in the P3b between correctly detected rare stimuli (solid line) and correctly ignored rare stimuli (broken line) when these neocerebellar patients were required to rapidly shift their focus of attention. In contrast, within even the shortest time interval following a cue to shift attention to a different perceptual domain, normal children displayed a large P3b response to targets in that new perceptual focus, but no P3b to correctly ignored rare stimuli in the old perceptual focus. Thus, following cues to shift, normal children were able to very quickly establish a new selective focus of attention and selective P3b enhancement, but children with neocerebellar damage were not. Calibration bars represent ±5 μV; total epoch = 1310 msec. (Adapted from Akshoomoff and Courchesne 1994).

new focus of selective attention. The Sd potential is absent in neocerebellar-lesioned patients and autistic patients.

NEW STUDIES OF ORIENTING ATTENTION AND THE CEREBELLUM

In addition to studies of shifting attention, a new set of experiments using the Posner paradigm (see Fig. 3, top; Posner et al. 1984) demonstrated neocerebellar involvement in a different domain of attention, the automatic orienting of attention (Townsend et al. 1996a). Although the Posner paradigm has been much studied and is familiar to most cognitive neuroscientists, use of one feature in its design has been overlooked, namely, the initial step of how quickly the spatial cue orients attention. In a typical Posner paradigm, on 80% of the trials, the cue correctly indicates the future location of the target stimulus (i.e., "valid" trial). It is therefore clear that the more quickly attention is oriented to the cue at this location, the more rapid the detection of and reaction time to a target at that location. The converse is also true; the more slowly attention is oriented, the slower the RT. An index of the speed of orienting attention (the Orienting Effect), then, compares the RT to short cue-to-target delays with the RT to long cue-to-target delays on valid trials (Fig. 3). Using this measure, Townsend et al. (1996a) demonstrated that patients with acquired focal neocerebellar lesions and autistic patients were substantially slower to orient attention than patients with focal frontal lesions, patients with parietal lesions, patients with developmental language delay (DLD), patients with attention-deficit/hyperactivity disorder (ADHD), normal children, and normal adults (Fig. 3).

These findings have been confirmed and extended by a new motor-free

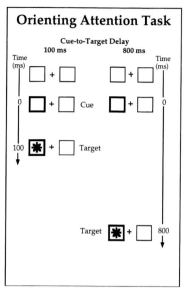

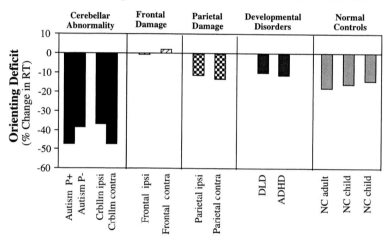

Figure 3: (*Top*) Diagram of one version of the Posner visual spatial attention task. Trials begin with the basic visual display, followed by the cue onset (box brightening) and 100 or 800 msec later by the target onset (asterisk). Subjects press a button in response to the target stimulus (Adapted from Townsend et al. 1996a). (*Bottom*) On such Posner tasks, patients with acquired neocerebellar lesions and autistic patients are slower to orient visual attention than patients with cerebral lesions, ADHD, DLD, or three groups of normal control subjects. The Orienting Effect is a measure that compares the RT to the short 100 msec cue-to-target delays with the RT to the long 800 msec cue-to-target delays on correctly cued trials [(RT at 800 msec) minus (RT at 100 msec) divided by (RT at 800 msec) times 100%]. By this measure larger negative scores mean a larger Orienting Deficit with the short 100 msec cue-to-target delay. P+ is data from autistic patients with MRI evidence of parietal volume loss; P– is data from autistic patients with no parietal volume loss. Data from patients with acquired neocerebellar lesions were from Townsend et al. (1996a); data from patients with unilateral parietal damage were from Posner et al. (1984); data from five patients with unilateral frontal lobe damage from pilot studies were adapted from Townsend, Courchesne, and colleagues (Courchesne 1995); adult normal control subjects from Townsend et al. (1996a); children with ADHD and normal control children from Swanson et al. (1991); normal controls and children with DLD from Nichols et al. (1995). (Adapted from Townsend et al. 1996a).

design in which accuracy of the attention-dependent discrimination, and not speed of the motor response, serves as an index of orienting attention effects (Townsend et al. 1996b).

In sum, studies of patients with neocerebellar damage have shown that such damage does slow the initiation or decrease the accuracy of operations that mobilize changes in attention, but it does not eliminate such operations.

NEW NEUROIMAGING EVIDENCE

The relatively recent arrival of fMRI brought with it the possibility of investigating the role of the healthy, normal cerebellum in attention. The first fMRI study to do this (Allen et al. 1997) was aimed at addressing two questions: (1) Is the cerebellum involved in attention operations that do not utilize the motor system; and (2) if there is such cerebellar involvement in attention, is it localized to the same region(s) involved in movement, or is it differentially localized within the cerebellum? This study employed three tasks. During an Attention task, circles, squares, or triangles in red, green, or blue were presented one at a time at a central fixation point. subjects silently counted target stimuli (squares or red shapes) within a predetermined visual dimension (form or color). Thus, this task required attention in the absence of a motor response. During a Motor task, subjects repeatedly executed a self-paced, right-hand movement in the absence of visual stimulation. This movement was then used in a final, attention-with-motor task, in which subjects responded to each target using the right-hand movement rather than silently counting target stimuli. To control for visual sensory stimulation, activation during both the attention and attention-with-motor tasks was compared with activation during passive visual stimulation, during which subjects observed the same set of visual stimuli but did not selectively attend or respond to targets. The motor task was compared with rest.

Thus, this study was designed to dissociate the involvement of the cerebellum in attention from its involvement in motor output and sensory input. such a dissociation was essential. The cerebellum has long been considered a motor control structure, and it receives input from a variety of sensory systems. Therefore, a crucial component of experiments attempting to demonstrate that its role extends beyond the motor and sensory domains is the careful control for motor and sensory activation. The Attention task controlled for motor output by employing a silent counting response involving neither the planning, the preparing, nor the executing of overt movements. As all stimuli were presented at a single spatial location in the center of foveal vision, eye movement activation was not predicted to occur. moreover, previous work would predict that had eye movements occurred, the resulting activation would have been observed in the cerebellar vermis (Petit et al. 1996), a region not activated during the Attention task. All areas that were active during the Attention task were also active during the Attention-with-Motor task, which did not employ silent counting, indicating that silent counting did not add to the results. Furthermore, when four subjects were instructed to silently count from 1 to 10 repeatedly in the absence of any visual stimuli, no cerebellar activation was observed within the most common and prominent site of attention-related activation. Finally, as visual sensory stimulation was the baseline control condition to which the two attention tasks were compared, this study controlled for sensory input as well.

By employing the above controls, the cerebellum was shown to be involved in selective attention operations independent of its involvement in motor output and sensory input. This involvement was dissociated neuroanatomically from cerebellar involvement in motor operations, with motor output activating the right anterior cerebellum and attention activating the superior posterior cerebellum, most prominently on the left (Fig. 4). Moreover, there was a sharp distinction between the manner in which motor output and attention activated these separate cerebellar regions. at the onset of the motor task, which was performed without any visual sensory stimulation, there was a transient increase in activation in the Attention hotspot, that is, in each subject, the maximally activated voxel in the left superior posterior cerebellum during the Attention task (Fig. 5A). This pattern of activation suggests that the initiation of the required motor output involved some degree of attention, but sustaining these simple actions did not. In contrast, during the Attention task, which was performed in the absence of motor planning or execution, there was no increase in activation in the Motor hot spot, that is, in each subject, the maximally activated voxel in the right anterior cerebellum during the Motor task (Fig. 5B). This suggests that neither the initiation nor the sustained execution of the Attention task required the use of cerebellar regions most involved in the Motor task. These results emphasize the functional independence of cerebellar involvement in attention; motor activation required attention, but attention activated the cerebellum irrespective of visual sensory input or motor output (Allen et al. 1997).

Recently, a separate group replicated the shifting attention paradigm developed in our laboratory (Akshoomoff and Courchesne 1994) and used

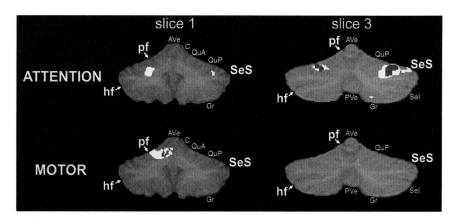

Figure 4: Functional maps demonstrating the most common sites of activation across subjects overlaid on averaged coronal anatomical images of the most anterior slice position (slice 1) and the most posterior slice position (slice 3) of the cerebellum analyzed. (Red) Overlap of three or more subjects; (yellow) any two subjects. During the Attention task, the most common site of activation was in the left superior posterior cerebellum [the posterior portion of the quadrangular lobule (QuP) and the superior portion of the semilunar lobule (SeS); approximate Talairach coordinates of center of mass ($x = -37$, $y = -63$, $z = -22$)]. During the Motor task, the most common site was in the right anterior cerebellum [the anterior portion of the quadrangular lobule (QuA), the central lobule (C), and the anterior vermis (AVe); approximate Talairach coordinates of center of mass ($x = 7$, $y = -51$, $z = -12$)]. (pf) Primary fissure; (hf) horizontal fissure; (PVe) posterior vermis; (SeI) superior portion of the semilunar lobule; (Gr) gracile lobule). (Adapted from Allen et al. 1997).

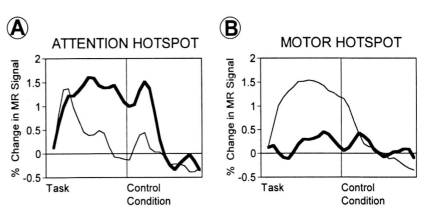

Figure 5: Intertask comparisons within averaged Attention (A) and Motor (B) Hot spots. For each hotspot, the time course signal data for each subject were averaged, collapsed across the four cycles between task and control conditions, and plotted in terms of percent change in MR signal. (Thick line) Attention task activation; (thin line) Motor task activation. (Adapted from Allen et al. 1997).

it to investigate with fMRI the role of the normal cerebellum in shifting attention (T.H. Le and X. Hu, unpubl.). These investigators reported that compared to sustained attention on a single dimension within the visual modality (e.g., color), shifts of attention between different dimensions induced activation in the right lateral cerebellum in all subjects and the ventral dentate nucleus in nearly half of subjects. This demonstration confirmed the predictions of previous neurobehavioral and neurophysiological studies: The healthy, normal cerebellum is involved in the dynamic control of attention.

ANIMAL STUDIES

In the theory of cerebellar preparatory function, the cerebellum predicts what is likely to happen next and what neural network conditions may therefore be needed, and then it signals preparatory changes in neural responsiveness of those networks expected to be needed in upcoming moments. Animal studies show that the cerebellum is in a position to be involved in specific as well as global preparatory processes.

First, animal studies show cerebellar involvement in arousal, alerting, and attention. In the report of the discovery of the reticular activating system (RAS), cerebellar stimulation was found to modulate the RAS response (Moruzzi and Magoun 1949), and subsequent experiments found that cerebellar stimulation triggers behavioral and parietal EEG alerting responses in primates (e.g., Siegel and Wepsic 1974). The size of the cerebellar molecular layer in normal rats is positively correlated with the degree of attention to novelty (Anderson 1994). Also, the cerebellum has connections with brain stem, thalamic, and cerebral systems involved in attention (for review, see Akshoomoff and Courchesne 1992; Courchesne et al. 1994).

Second, when the cerebellum in fish and mammals is activated in advance of sensory information, neural responsiveness to subsequent sensory events is altered in nonmotor brain stem, thalamic, cerebral, and hippocampal sites. Such effects have been documented for visual, auditory, somatosensory, and nociceptive stimuli (e.g., Newman and Reza 1979; Crispino and Bullock 1984; Liu et al. 1993). For example, stimulation of cerebellar vermian lobules VI and VII in awake but nonbehaving rats modulates superior colliculus (Crispino and Bullock 1984) and hippocampus (Newman and Reza 1979) responses to a sensory stimulus if the vermis stimulation occurs in advance of the sensory stimulus. The effectiveness of cerebellar stimulation varies with time and is described by a response modulation tuning curve with the optimal peak being at 50 msec *in advance of the subsequent sensory stimulus* (Fig. 6). This modulation is independent of motor involvement.

Third, when the cerebellum is activated in advance of sensory information, there is a signal-to-noise enhancement of subsequent sensory responses in brain stem, thalamic, and cerebral sites. This is a crucial property of cerebellar functioning because it allows the optimization of neural conditions for the acquisition of sensory information. For instance, when background luminance is sufficient to reduce to noise levels the colliculus response to a flash stimulus, stimulation of cerebellar vermian lobules VI and VII in rats causes the colliculus response to that flash to emerge above noise levels *if the cerebellar stimulation occurs in*

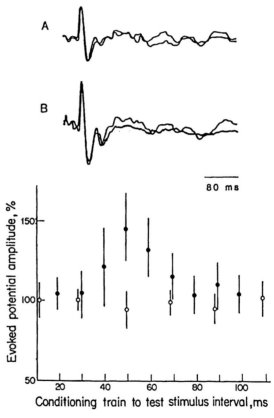

Figure 6: (*Top*) Effect of cerebellar stimulation on the flash-evoked potential in the superior colliculus in the rat. (A) Control response; (B) after cerebellar stimulation applied 50 msec before flash. Superimposed averages, 64 sweeps each. (*Bottom*) Relationship between the relative amplitude of the superior colliculus flash-evoked response and the interval between cerebellar stimulus and flash in the rat. (○) Control responses; (●) responses after cerebellar stimulation. Bars represent the whole range of responses. Each point is the average of four experiments of 64 stimulus presentations each. (Adapted, with permission, from Crispino and Bullock 1984).

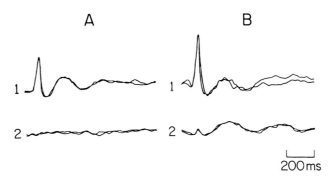

Figure 7: Combined effects of background illumination, flash, and cerebellar stimulation on superior colliculus evoked potentials. (*A, 1*) Control response to flash; (*A, 2*) abolition of such response by addition of background light; (*B, 1*) response to flash after cerebellar stimulation; (*B, 2*) response to flash, under background illumination, after cerebellar stimulation. Superimposed averages, 64 stimulus presentations. (Adapted, with permission, from Crispino and Bullock 1984).

advance of the visual stimulus (Crispino and Bullock 1984) (Fig. 7). This improvement in signal-to-noise is independent of motor involvement.

This cerebellar property may help explain the poor performance of patients with cerebellar lesions when they must rapidly orient attention to detect a brief flash of light. In the Posner paradigm (Fig. 3, *top*), the box brightening serves as a conditioned stimulus to a subsequent flash of light. We suggest that the normal cerebellum facilitates detection performance by learning this association and using the box brightening as a signal to enhance signal-to-noise conditions (e.g., in the superior colliculus) in preparation for the subsequent flash. As in the Crispino and Bullock experiments, this cerebellar preparatory response enhancement period may be brief. Therefore, lesions of the cerebellum would be expected to impair this preparatory enhancement mechanism and so, impair detection of the subsequent flash. Such impaired detection abilities have been reported in patients with neocerebellar lesions, as discussed above (Townsend et al. 1996a,b) (Fig. 3).

Fourth, the cerebellum appears to be able to effect global preparatory actions, in addition to its ability to effect sensory modality specific changes. Stimulation of the fastigial nucleus in rats, cats, rabbits, and primates increases rCBF throughout the brain, and in the rat, such rCBF increases in cerebral cortex, hippocampus, and amygdala were independent of changes in metabolism (Nakai et al. 1983). The investigators hypothesized that the fastigial nucleus "may elicit preparatory changes in rCBF to increase flow of substrate for cerebral metabolism, in anticipation of the possibility of a more widespread increase in metabolism elicited by the (subsequent) actual performance of behaviors."

Fifth, the effectiveness of preparatory cerebellar activation lasts for variable lengths of time, perhaps thereby allowing the cerebellum to help smoothly integrate rapid moment-to-moment response demands within a larger and longer-term goal-oriented framework. At one extreme, the cerebellar preparatory response enhancement period lasts on the order of 50 msec (Crispino and Bullock 1984) (Fig. 6). At the other extreme, following the cessation of cerebellar stimulation, the subsequent changes to pain thresholds in primates (Siegel and Wepsic 1974), to neural spike activity in parafasciulus thalami evoked by noxious stimulation in rats (Liu et al. 1993), and to somatosensory cortex responses in humans (Snider and Mitra 1970) last for many minutes.

The preparatory role of the cerebellum encompasses motor functions as well as nonmotor sensory, attention, or even cerebrovascular functions. For instance, in the monkey, repeated perturbation of a handheld object has been shown to evoke preparatory increases in grip force that occur in anticipation of perturbation. These changes in grip are accompanied by the modulation of simple spike discharges in cerebellar Purkinje cells (Dugas and Smith 1992). Likewise, changes in amplitude of the vestibulo-ocular reflex (VOR) have been shown to occur in anticipation of changes in vergence angle, and transient Purkinje cell discharges in association with such changes in vergence occurred "early enough to drive the changes in VOR amplitude" (Snyder et al. 1992).

Using their own model of cerebellar function, Coenen and Sejnowski (1996) computationally reproduced results similar to those of Snyder et al. (1992). These investigators also view the cerebellum as a structure that learns to predict neural events, and "therefore is particularly successful in anticipating the temporal sequences of events experienced repeatedly" (Coenen and Sejnowski 1996). In this model, the deep cerebellar nuclei are specialized to encode immediate or short-term predictions, whereas the cerebellar cortex, because it receives much more complete neural contextual information, can inhibit these short-term predictions of the cerebellar nuclei and replace them with longer-term, higher-quality predictions. In addition, in this model, the inferior olive integrates inhibitory input from the deep cerebellar nuclei and excitatory sensory inputs to compute a prediction error that is then reported back to the cerebellar cortex. Inhibitory input is also used to modulate the synchronous firing of olivary neurons and thus change the number of climbing fibers reporting the prediction error, which in turn changes the number of Purkinje cells that will be utilized to learn a given prediction. Thus, in this model the cerebellum serves a predictive function that is "under the regulatory control of the inferior olive" (Coenen and Sejnowski 1996).

Another recent hypothesis is that the cerebellum guides movements so as to improve the "quality of sensory information" received (Bower and Kassel 1990; Gao et al. 1996). According to this model, during active tactile exploration of novel objects or environments, or during a search for a known and expected tactile stimulus, the cerebellum uses movement as a tool for repositioning tactile receptors so as to "improve the efficiency of sensory processing (*) by the rest of the nervous system." If we insert into this quote at the * the phrase "of subsequent stimuli," then we have a more complete account of the preparatory purpose of cerebellar signals that affect the repositioning of tactile receptors. That is, such motoric repositioning is a *preparatory action* occurring in advance of the next moment of sensory acquisition and serving to set conditions needed for that predicted moment of acquisition. In this way, the cerebellum may use movement to accomplish goals that are comparable as well as complementary to those achieved by the nonmotor repositioning of the focus of attention. Thus, the particular hypothesis that the cerebellum "adjusts" movement to improve acquisition of anticipated tactile information is fully consistent with the general theory of cerebellar function presented herein.

Additionally, because the cerebellum can improve the quality of sensory information *without* any movement (Fig. 7), the common denominator of functional importance is not motor control alone but, rather, preparatory control involving a wide variety of mechanisms (e.g., motor, sensory, attention, cerebrovascular).

For the cerebellum to implement preparatory responses, it must signal activations according to predictions about what is about to happen. This requires prior learning of temporally ordered sequences of endogenously derived and exogenously derived neural activites. A wealth of evidence from PET and fMRI studies as well as animal studies supports a role for the cerebellum in such learning.

Functional Neuroimaging Evidence of Cerebellar Involvement in Learning

CONDITIONAL ASSOCIATIVE LEARNING

The idea that the cerebellum might be involved in associative learning dates back decades (e.g., Brogden and Gantt 1942). In 1971, Asdourian and Preston showed that cerebellar stimulation can serve as a conditioned stimulus. They suggested that "if an animal can utilize activity originating in the cerebellum as a signal for some subsequent event, then the traditional role assigned to the cerebellum as a silent coordinator of sensory-motor functions will need reevaluation."

Presently, the importance of the cerebellum to associative learning has been well documented (McCormick and Thompson 1984; Krupa, Thompson and Thompson 1993) and confirmed independently with PET by three groups (Molchan et al. 1994; Logan and Grafton 1995; Blaxton et al. 1996). All three examined activation during the acquisition of an association between a tone and an air puff to the right cornea. The first such study measured regional cerebral blood flow (rCBF) during three conditions: (1) unpaired presentations of the tone and air puff; (2) conditioning trials during which the tone preceded and terminated with the air puff; and (3) extinction trials during which the tone was presented in the absence of an air puff (Molchan et al. 1994). This study reported that relative to the unpaired and extinction conditions, conditioning trials resulted in significantly decreased activation in the right cerebellar cortex. In the second study, which measured relative cerebral glucose metabolism, Logan and Grafton (1995) also compared activity during the presentation of random, unpaired tones and air puffs to that during paired conditioning trials. They reported that relative to the unpaired condition, activity increases in the paired condition were seen in several bilateral and midline cerebellar sites. They also showed that activity in these regions correlated positively with increases in subjects' conditioned responses (i.e., activation correlated positively with the degree of associative learning that occurred). Finally, Blaxton et al. (1996) measured rCBF during visual fixation, uncorrelated presentations of air puffs and tones, and four separate paired conditioning trials. This design allowed a more detailed test of how activation changes over the course of acquisition. Like Logan and Grafton (1995), these investigators showed that activity increased bilaterally in the cerebellum as the number of conditioned responses produced by their subjects increased.

The proposed theory of preparatory learning predicts that the initial stages of learning in the cerebellum will be accompanied by increases in cerebellar activation. In two of the associative learning studies reviewed here (Logan and Grafton 1995; Blaxton et al. 1996), this pattern was observed. However, in one study, no increased cerebellar activation was observed during learning, the only cerebellar finding being decreased right cerebellar cortex activation during conditioning trials (Molchan et al. 1994). What might account for this inconsistency?

A crucial aspect of any functional neuroimaging study is the careful control for unintended differences between contrasting conditions that may introduce alternate explanations for the observed changes in activation. In the case of a classical conditioning experiment in which the standard comparison is between activation induced by unpaired stimuli versus that induced by learning the association between unconditioned and conditioned stimuli, it is crucial that subjects have an equal amount of experience with both stimulus conditions prior to image acquisition. If not, the study may be more an investigation of the brain response to a novel stimulus situation rather than the response to learning conditional associations. There are indications that the cerebellum is involved in the response to novelty (Anderson 1994). Moreover, the proposed theory of cerebellar function predicts that a novel stimulus situation will elicit cerebellar activation reflecting cerebellar attempts to learn and make sense of the new environment such that it might ultimately prepare neural systems required to respond to the novel stimulus demands. Therefore, control for novelty is essential. In both studies demonstrating an increase in activation with conditioning (Logan and Grafton 1995; Blaxton et al. 1996), subjects received a number of tone and air puff presentations prior to the onset of image acquisition for both the unpaired and the conditioning trials. In contrast, Molchan et al. (1994), who reported an overall decrease in cerebellar activation during the conditioning trials, initiated the unpaired and extinction trials only 30 sec prior to the onset of the PET scan. However, for the conditioning trials, subjects had 18 min of prescan experience with the stimulus pairs for the first scan and 30 min of prescan experience for the second. Thus, prior to the unpaired and extinction conditions, subjects were exposed to far fewer trials than they were prior the learning condition. This discrepancy points to the possibility that the observed cerebellar deactivation during learning was attributable to the novelty of the unpaired and extinction conditions relative to the learning conditions. Increased activation due to such novelty may have obscured any changes due to the factor of interest, associative learning.

SEMANTIC ASSOCIATION LEARNING

In the first PET report of cerebellar activation during a cognitive task, Petersen et al. (1989) described activity increases in the right inferior lateral cerebellum when subjects spoke verbs appropriate to nouns that they saw or heard. That same group later used PET to image the effects of practice on rCBF during this verb generation task (Raichle et al. 1994). These investigators found that ~15 min of practice resulted in a significant reduction in activation in the right cerebellar hemisphere to the point that it was comparable to activation seen during simple noun repetition, whereas generation of verbs to a novel set of nouns caused activation to return to prepractice levels (Fig. 8). The investigators interpreted these results as indicating that two separate neural pathways subserve performance of the verb generation task. One pathway operates when the task is novel, and the other operates when the task is in the learned or automatic state. Raichle et al. (1994) proposed that involvement of the right cerebellar hemisphere in the task when it was novel, but not when it was learned, serves as evidence that the right cerebellar hemisphere is

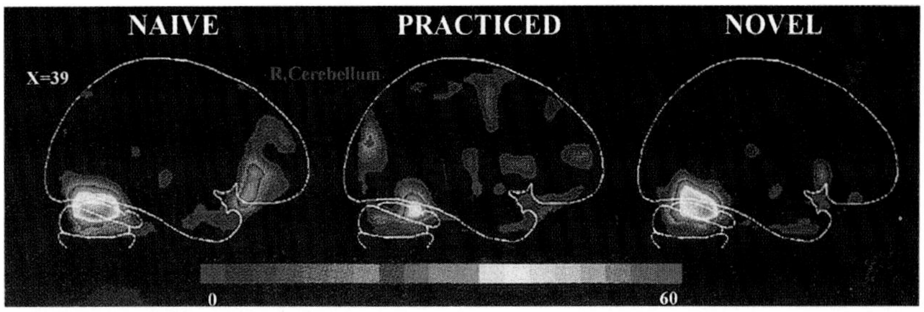

Figure 8: PET difference images show areas of *increased* blood flow in the right cerebellar hemisphere when subjects spoke a verb appropriate to a visually presented noun relative to simply speaking aloud the visually presented noun itself. The color scale is a linear scale of normalized radioactive counts, with maximum and minimum as shown. Brain outlines were traced from the stereotaxic atlas of Talairach and Tournoux (1988) and represent a sagittal slice position 39 mm to the right of midline. The three images represent the three conditions of the experiment (i.e., naive, practiced, and novel). Note the reduced right cerebellar activity in the practiced condition (*center*) that reappeared in the novel condition (*right*). (Adapted, with permission, from Raichle et al. 1994).

"of critical importance in practice-related learning and the detection of errors in a variety of tasks involving complex nonmotor processing."

The study of Raichle et al. (1994) provides a strong demonstration of cerebellar involvement in learning outside of the pure motor domain. It is important to emphasize here that cerebellar hemisphere activity did not disappear completely when the verb generation task was in the practiced state; it simply reduced. Although the investigators suggest that this represents a shift from the use of one neural pathway to the use of another, the persistent cerebellar activity might also reflect a learning-dependent modulation in the degree to which a single pathway involving the cerebellum is being utilized. We suggest that during the "naive" performance of the verb generation task, increased cerebellar activation reflects the initial attempts at learning semantic associations and preparing neural systems involved in the required response while cerebellar activation that remained after the task had been practiced reflects more refined and efficient cerebellar preparatory learning continuing to aid neural systems required to perform this task.

PROCEDURAL LEARNING

Several investigators have examined cerebellar involvement in procedural learning, that is, learning to perform a motor action in response to a particular sensory input. For instance, the question of which neuroanatomic regions show changes in rCBF during learning of the pursuit rotor task was investigated with PET by Grafton et al. (1994). In this task, subjects are instructed to maintain contact between the tip of a metal stylus held in the right hand and a 2-cm target located on a rotating disc. The key to improving performance on this task is to learn to predict the motion of the target and the necessary location of the hand to maintain contact. Thus, as it requires the learning of predictions, this task should be especially effective in eliciting cerebellar activation. In fact, Grafton et al. (1994) showed that during the early stages of learning this task, significant increases in activation across trials were observed in the anterior cerebellum ipsilateral to the moving right hand. Furthermore, when correlating the actual rate of performance improvement with changes in activation, the left anterior cerebellum showed a positive correlation. Together, these findings were seen as "strong support for the widely held idea that the cerebellum is critically involved in motor learning" (Grafton et al. 1994).

The structures involved in "cognitive-motor learning" were examined with fMRI (S.L. Rao, D.L. Harrington, K.Y. Haaland, J.A. Bobholz, J.R. Binder, T.A. Hameke, J.A. Frost, B.M. Myklebust, R.D. Jacobson, P.A. Bandettini, and J.S. Hyde, unpubl.). In their task, an adaptation of the serial reaction time (SRT) task, subjects executed one of four possible key presses with the right index finger in response to a visual stimulus appearing in one of four possible spatial locations. Subjects performed multiple blocks of trials consisting of a repeating 12-element sequence, that is, a nonrandom sequence that could be learned. Thus, this study provides a test of one aspect (i.e., sequence learning) of the theory of cerebellar function proposed herein. During the first five blocks of trials, subjects showed evidence of learning the sequence as indexed by decreases in reaction time. This learning was paralleled by significantly increased activation primarily in the left lateral cerebellum as compared to activation during rest. The investigators argued that this pattern of activation demonstrates the importance of the left lateral cerebellum to cognitive-motor sequence learning, and they also pointed out that the left lateralization of the activity modulations suggests that the effect was unrelated to the motor aspects of the task. In fact, this lateralization is consistent with the involvement of the cerebellum in learning sequences to aid performance of the task at hand. In this case, the task at hand has a significant visual spatial attention component, a type of task thought to involve the right cerebral hemisphere more than the left and, hence, the left cerebellar hemisphere more than the right.

In another test of cerebellar involevment in sequence learning, Jenkins et al. (1994) imaged with PET while subjects learned by trial and error a sequence of key presses in response to a pacing tone. Subjects were scanned during rest; during learning of a new, unfamiliar sequence; and during a prelearned, highly practiced sequence. During the unfamiliar sequence, activation foci were seen bilaterally in the cerebellar cortex and the cerebellar nuclei in addition to the vermis, whereas during the prelearned task, only the vermis and nuclei were active. Moreover, in terms of percent signal change, the magnitude of activation during the unfamiliar sequence was greater than that during the prelearned task. So, as in Raichle et al. (1994), this study indicated that the transition from naive to highly practiced task performance is accompanied by reduced activation in the cerebellum. This study also suggests that such reductions are more prominent in cerebellar cortex than they are in the cerebellar nuclei. The investigators interpreted their results as being "compatible with the hypothesis that the cerebellum plays some role in the process by which learned tasks become automatic" (Jenkins et al. 1994).

Finally, Flament et al. (1996) used fMRI to measure cerebellar involvement in three different conditions of a tracking task requiring subjects to use joystick movements to superimpose a cursor on eight targets appearing one at a time at eight different spatial locations. In the "standard" condition, joystick movement corresponded exactly with movement of the cursor. In the "random" condition, there was no predictable relationship between joystick and cursor movement. Finally, in the "reversed" condition, the relationship was constant but reversed

Functional Neuroimaging of Cerebellar Prediction and Preparatory Learning

The proposed theory of cerebellar function suggests a reinterpretation of a wide variety of PET and fMRI results while also guiding the formulation of hypotheses about how new tasks might be expected to activate the cerebellum. For it to implement its prime preparatory function, the cerebellum must first learn sequences of temporally ordered, multidimensional, exogenously and/or endogenously derived neural activities. This is no small task. The volume of information coming into the cerebellum far exceeds that which ultimately leaves. In fact, the ratio of cerebellar afferents to efferents has been estimated to be ~40:1 (Carpenter 1991). This massive converging input to the cerebellum is hypothesized to be reflected in marked increases in functional activation as attempts at learning and integrating this input commence. Then, as learning progresses, cerebellar activation will decrease, with the rate, degree and regional extent of decrease being a function of both the difficulty and complexity of the task at hand, and the amount of explicit knowledge subjects have of the task goal (Fig. 10).

Although it is decreasing, activation specific to learning will not disappear. Rather, it is hypothesized to continue changing as long as changes in the sequence to be learned occur. What is initially a large region of activation reflecting the initial attempts at learning and generating preparatory output will ultimately become a stable area of activity reflecting the regional extent of neural tissue required to perform a particular motor or mental action whenever it is encountered in the future. At one extreme, if the sequence is random and thus constantly changing, cerebellar activation will be sustained at a relatively constant magnitude for the duration of the task as the cerebellum attempts (yet fails) to discern a meaningful and predictive pattern. In contrast, if the sequence is very repetitive and thus predictive, the cerebellum will learn quickly, and this will be reflected in a rapid decrease in learning-related cerebellar activation with experience. In the Attention task used by Allen et al. (1997), the random relationship between the target stimuli to which subjects were to respond and the nontarget stimuli that preceded them precluded the cerebellum from learning a meaningful, predictive sequence. Despite the lack of any predictive relationship between events, we argue that the cerebellum searched for one, and activation was sustained through the duration of the task as the cerebellum attempted to learn. Figure 5A shows that during the Attention task, activation was sustained at a relatively constant magnitude in the Attention Hot spot. In contrast, activation in this location showed a strikingly different pattern during a task involving simple repetitive finger movements. At the onset of this Motor task, there was a transient increase in activation in the Attention Hot spot, which then quickly returned to baseline long before the end of the task (Fig. 5A). One interpretation of this pattern is that the repetitive nature of the task, in which events occurring in the present (the motor actions) were predicted by those occurring in the past (the same motor actions), enabled the cerebellum to learn quickly. As the nature of the task went unchanged once initiated and no new learning was required, cerebellar activation in this particular region waned. This transient activation may represent the amount of neural activity from this particular cerebellar region necessary to generate a preparatory response for systems involved in performing these simple motor actions each time they are to be encountered.

Interpreting the results from the experiment of Allen et al. (1997) in light of the proposed theory of preparatory learning also provides an explanation for certain neurobehavioral results from our laboratory (see above, Early Studies of Shifting Attention and the Cerebellum). Although the sustained attention task consistently activates the cerebellum, patients with cerebellar lesions are unimpaired when performing such a task (Akshoomoff and Courchesne 1992, 1994; Courchesne et al. 1994). Although this effect might seem intuitively paradoxical, it makes perfect sense when reconsidered in the context of a cerebellar preparatory function. As the sequence of events in this task is randomly ordered and thus unpredictable, the cerebellum is unable to learn and provide a preparatory service for systems involved in detecting and responding to target stimuli. Although the cerebellum is active in its attempts to learn the sequence, its failure to learn precludes it from providing effective aid to these regions of the CNS. Thus, when a particular sequence to be learned is completely random, the normal cerebellum has less of an advantage over the damaged cerebellum, and successful performance is not significantly impeded by cerebellar lesions. On the other hand, performance of the shifting attention task, a task that also activates the cerebellar cortex in addition to the dentate nucleus (T.H. Le and X. Hu, unpubl.), *is* impeded by cerebellar lesions (Akshoomoff and Courchesne 1992, 1994; Courchesne et al. 1994). In this task, the appearance of target stimuli cues subjects to attend to a new stimulus dimension and prepare for a target, thus entailing a predictive relationship between stimuli that the cerebellum is able to learn. In turn, the cerebellum can provide useful preparatory output to neural systems required to detect and respond to the next target. We suggest that an impairment of learning and implementing such a preparatory output may be the culprit behind impaired performance of the shifting attention task in patients with lesions to the neocerebellum.

Activation specific to learning is hypothesized to occur in the cerebellar cortex, the point of convergence for massive inputs to the cerebellum. Further predictions about within-cortex activation can be made based on the purpose for which the cerebellum is learning a particular sequence. The majority of cerebral input to the human cerebellum comes from the contralateral hemisphere via cerebro-ponto-cerebellar projections innervating the posterior lobe of the cerebellum (Ito 1984). Thus, if learning in the cerebellum is going to aid the efficient performance of a language task or a task typically performed with a verbal strategy, activation will be more prominent in the right posterior cerebellar hemisphere, as shown now by several functional neuroimaging studies employing language tasks (e.g., Petersen et al. 1989; Raichle et al. 1994; Martin et al. 1995; E. Artiges, M.J. Giraud, B. Mazoyer, C. Trichard, L. Mallet, H. de la Caffiniere, M. Verdys, A.M. Syrota, and J.L. Martinot, unpubl.) (Fig. 8). In contrast, if the cerebellum is attempting to learn a sequence to facilitate performance of a sustained, selective attention task, a type of task thought to involve right prefrontal cortex more than left (Stuss et al. 1995), cerebellar activation is expected to be more prominent in the left posterior cerebellum, as demonstrated by Allen et al. (1997) (Fig. 4).

Once the cerebellum has begun to learn a particular sequence, it will begin to predict what is about to happen and prepare other neural systems accordingly. This preparatory output will begin as soon as the cerebellum has learned enough to generate an approximation of the sequence in an attempt to prepare other neural systems. Initially, this preparation will not be perfect, and the "error," that is, the discrepancy between the form, degree, and timing of the preparation that is required and that which has been achieved, will be fed back to the cerebellum, aiding further attempts to learn the sequence. Such adaptive adjustments will improve the quality of a particular preparatory response. As greater precision in prediction and preparation is achieved, activation reflecting learning in the cerebellar cortex will decrease. This change may manifest as a shift from widespread activation to more focal sites of activity, as seen, for instance, in Raichle et al. (1994) (Fig. 8).

What would be the functional activation "signature" of prediction and preparation by the cerebellum? With respect to a particular sequence, this form of activation would not be expected to be apparent during the earliest attempts at learning. However, shortly after learning begins, early attempts at a preparatory output will be made, and this will be reflected in an emergence of related activation. Such activation may then decrease as the preparatory response becomes more efficient, but it will continue throughout performance of the task at hand, and it will re-emerge whenever the particular context is encountered in the future. In the functional neuroimaging studies reviewed here, cerebellar activation did not disappear once learning had occurred. For instance, Flament et al. (1996) reported that over the course of learning a "visuomotor dissociation task," activation in "both intermediate and lateral cerebellar cortex" reduced but did not disappear (Fig. 9). We suggest that such cerebellar activity that remains after learning has occurred may in fact reflect a combination of preparatory output from the cerebellum to systems involved in the task, and continued learning of adaptive adjustments to the quality of this output in response to feedback from such systems.

As the cerebellum provides preparatory output to other neural systems, activation is also hypothesized to emerge in the sites of cerebellar output, the deep cerebellar nuclei. In a subset of studies reviewed here, such nuclei activation did occur. For instance, Flament et al. (1996) and Jenkins et al. (1994) both reported that the areas in which activation was sustained after tasks had been learned included the cerebellar nuclei. Overall, however, cerebellar nuclei activations have not been a common occurrence in functional neuroimaging studies of the cerebellum, and the reason for this is not completely clear. One possibility is that to observe activation in the output nuclei by conventional design and analysis methods, a task must require continuous updating in the preparatory output emanating from the cerebellum. A possible example of such a task was employed by Gao et al. (1996), who explored with fMRI the role of the cerebellum in shape and texture discrimination. These investigators reported bilateral activations in the dentate nuclei during shape discrimination involving finger movements and during texture discriminations that involved no explicit finger movements, but not during finger movements alone. The discrimination tasks used a match-to-sample design in which subjects were presented with a sample shape or texture, followed by a new shape or texture, and were then required to determine whether the new item was identical to the sample. Thus, each stimulus presentation served as a predictor for a subsequent required discrimination. This continuous presentation of predictors may have driven the cerebellar output nuclei to remain active as they continuously prepared relevant efferents for the upcoming discriminative judgements.

Such continuous prediction and preparation may also explain the dentate nucleus activation observed in the fMRI study of shifting attention of T.H. Le and X. Hu (unpubl.). Unlike the sustained attention task employed by Allen et al. (1997), the shifting attention task enables the cerebellum to learn and generate a more precise and informative preparatory response. In this task, target stimuli cue subjects to shift attention to a new stimulus dimension and prepare for a new target. Thus, there is a predictive relationship between stimuli that the cerebellum should be able to learn quite readily. In turn, the learning of this relationship should allow the cerebellum to generate a preparatory output to neural systems required to detect and respond to upcoming target stimuli. We suggest that dentate activation in the Le and Hu study may in fact reflect such cerebellar preparatory output.

Activation in the cerebellar nuclei is hypothesized to continue throughout performance of the task at hand. However, with continued adaptive adjustments, the preparatory output is expected to approach a state of optimal quality, and as greater precision is achieved, activation reflecting such output may decrease. For instance, Friston et al. (1992) and Seitz et al. (1994) both imaged subjects during the learning of motor tasks having a relatively low level of difficulty and not involving the complex stimulus–response associations characteristic of procedural learning paradigms. During such tasks, the duration of the initial learning phase is likely brief, as is the amount of time that transpires prior to the onset of cerebellar preparatory output. Moreover, because such tasks are relatively simple and repetitive, once the initial preparatory responses are generated by the cerebellum, little more adaptation to the preparation is required before subjects reach a stage of optimal performance. Thus, during such tasks, preparatory output-related activation might actually show a decrease as the preparatory response becomes more efficient and precise. We suggest that the attenuation in activation of the cerebellar nuclei that was observed after the acquisition of motor skills in the studies of Friston et al. (1992) and Seitz et al. (1994) may in fact have been a reflection of such an optimization of preparatory output.

Whether the hypothesized patterns of cerebellar activation will be observed in any particular study is largely dependent on both the careful design of tasks and the thoughtful choice of analysis procedures. The proposed preparatory operation of the cerebellum is a dynamic process, constantly adapting to internal and external variation in an attempt to provide other neural systems with the most up-to-date preparatory information possible. Each phase of this dynamic operation is less likely to be detected in the context of the conventional subtraction or "box-car" correlation method of design and analysis, as these approaches are intended to detect simple dichotomous variation between two conditions, not the complex and dynamic patterns of change that are expected in light of the proposed theory.

Although it is dynamic with respect to a particular sequence, the proposed preparatory operation is also an ongoing function that persists as the cerebellum attempts to make sense of, or get the gist of, what is occurring at any given time. This may explain the fact that the cerebellum seems to be active in some way almost regardless of the task employed (see Table 1 for several examples). Up until now, methods of functional image analysis have almost always involved simple comparisons between conditions. Yet, if nearly every condition activates the cerebellum in some way, then conclusions about how any one task of interest activates the cerebellum are going to be limited by conventional design and analysis procedures. Thus, future tests of cerebellar function will require far more sophisticated design and statistical approaches to investigating functional activation changes in the cerebellum.

Conclusions

A long line of research investigating diverse species and diverse operations supports the theory of a prediction and preparation learning function for the cerebellum. Studies of fish, rats, rabbits, cats, monkeys, and humans using intracerebellar stimulation, focal lesion, neurophysiological, neurobehavioral, or neuroimaging methods show that the cerebellum can and does prepare/set a variety of internal conditions in advance of sensory events and neural operations. There is experimental evidence that cerebellar signal effects can alter neural responses to subsequent sensory stimuli that occur not only milliseconds later, but also seconds or minutes later. So, this cerebellar influence can be local and brief (e.g., signal-to-noise changes in the superior colliculus), global and last for seconds (e.g., rCBF changes), or even local and last for minutes (e.g., neural responsiveness in parafasciulus).

The observations that the cerebellum can trigger changes that can last seconds or minutes stand in contrast to the computational models and traditional experiments (e.g., eye-blink conditioning, VOR) that support cerebellar associative learning and predition time scales in the millisecond to hundreds of milliseconds range. This contrast highlights the need for new computational models of the cerebellum and new designs for experiments of cerebellar associative learning that will bridge this "temporal gap." Previous concepts of the cerebellum have tended to be derived from an unnecessarily narrow base of sensory and motor experimental paradigms and data. The explanatory power of such concepts may be correspondingly limited. For instance, how do motor control, motor learning, sensory tracking, error detection and correction, or timing hypotheses of the cerebellum account for the reciprocal cerebellar connctions with virtually all hypothalamic nuclei (Haines and Dietrichs 1987), for cerebellar modulation of parafasciular neuron responsiveness to noxious stimuli (Liu et al. 1993), or for cerebellar stimulation affecting rCBF? It is likely that more powerful explanatory concepts will emerge in the future as the paradigm and databases are broadened.

At present, a large body of neuroimaging and neurobehavioral evidence in humans shows cerebellar involvement in a remarkably diverse array of cognitive as well as noncognitive operations (e.g., Table 1). The current preparatory theory of the cerebellum helps to explain the involvement of the cerebellum in this wide array of operations; it helps to explain the functional importance of massive input from cerebral cortical association areas, particularly prefrontal, which may be crucial during initial, conscious-controlled stages in preparatory learning; and it moves us away from the "motor mind-set" that has stood for so long as a roadblock limiting our ability to fully comprehend the function of the cerebellum.

The fact that virtually any sensory, motor, or cognitive task can lead to cerebellar activation is a phenomenon that, like the question of cerebellar function itself, has either perplexed or been ignored by the majority of neuroscientists for many years. It is our hope that the theory proposed here will provide neuroscientists with a fresh viewpoint from which to reconsider cerebellar involvement in operations ranging from simple motor actions and cerebrovascular control to complex executive functions.

Acknowledgments

This work was supported by funds from the National Institutes of Neurological Disorders and Stroke (2-RO1-NS-19855) and the National Institute of Mental Health (1-RO1-MH-36840) awarded to E.C. and a McDonnell–Pew Graduate Fellowship in Cognitive Neuroscience awarded to G.A. We thank Olivier Cocnen and Pamela Moses for helpful comments on the manuscript.

References

Akshoomoff, N.A. and E. Courchesne. 1992. A new role of the cerebellum in cognitive operations. *Behav. Neurosci.* **106:** 731–738.

———. 1994. Intramodality shifting attention in children with damage to the cerebellum. *J. Cognit. Neurosci.* **6:** 388–399.

Akshoomoff, N.A., E. Courchesne, and J. Townsend. 1997. Attention coordination and anticipatory control. In *International review of neurobiology. The cerebellum and cognition* (ed. J.D. Schmahmann), Vol. 41, Academic Press, San Diego, CA. (In press.)

Allen, G., R.B. Buxton, E.C. Wong, and E. Courchesne. 1997. Attentional activation of the cerebellum independent of motor involvement. *Science* **275:** 1940–1943.

Anderson, B. 1994. The volume of the cerebellar molecular layer predicts attention to novelty in rats. *Brain Res.* **641:** 160–162.

Andreasen, N.C., D.S. O'Leary, S. Arndt, T. Cizadlo, R. Hurtig, K. Rezai, G.L. Watkins, L.L. Boles Ponto, and R.D. Hichwa. 1995a. Short-term and long-term verbal memory: A positron mission tomography study. *Proc. Natl. Acad. Sci.* **92:** 5111–5115.

Andreasen, N.C., D.S. O'Leary, T. Cizadlo, S. Arndt, K. Rezai, G.L. Watkins, L.L. Boles Ponto, and R.D. Hichwa. 1995b. Remembering the past: Two facets of episodic memory explored with positron emission tomography. *Am. J. Psychiatry* **152:** 1576–1585.

Asdourian, D. and R.J. Preston. 1971. Cerebellar stimulation as a CS. *Physiol. Behav.* **6:** 235–239.

Awh, E., E.E. Smith, and J. Jonides. 1995. Human rehearsal processes and the frontal lobes: PET evidence. *Ann. N.Y. Acad. Sci.* **1769:** 97–117.

Bava, A., T. Manzoni, and A. Urbano. 1966. Cerebellar influences on neuronal elements of thalamic somatosensory relay-nuclei. *Arch. Sci. Biol.* **50:** 181–204.

Berman, K.F., J.L. Ostrem, C. Randolph, J. Gold, T.E. Goldberg, R. Coppola, R.E. Carson, P. Herscovitch, and D.R. Weinberger. 1995. Physiological activation of a cortical network during performance of the Wisconsin card sorting test: A positron emission tomography study. *Neuropsychologia* **33:** 1027–1046.

Blaxton, T.A., T.A. Zeffiro, J.D.E. Gabrieli, S.Y. Bookheimer, M.C. Carrillo, W.H. Theodore, and J.F. Disterhoft. 1996. Functional mapping of human learning: A positron emission tomography activation study of eyeblink conditioning. *J. Neurosci.* **16:** 4032–4040.

Bower, J.M. and J. Kassel. 1990. Variability in tactile projection patterns to cerebellar folia crus IIA of the Norway rat. *J. Comp. Neurol.* **302:** 768–778.

Brogden, W.J. and W.H. Gantt. 1942. Intraneural conditioning: Cerebellar conditioned reflexes. *Arch. Neurol. Psychiatry* **48:** 437–455.

Carpenter, M.D. 1991. *Core text of neuroanatomy,* 4th ed. Williams & Wilkins, Baltimore, MD.

Ciesielski, K.T., E. Courchesne, and R. Elmasian. 1990. Effects of focused selective attention tasks on event-related potentials in autistic and normal individuals. *Electroencephalgr. Clin. Neurophysiol.* **75:** 207–220.

Coenen, O.J.M.D. and T.J. Sejnowski. 1996. Learning to make predictions in the cerebellum may explain the anticipatory modulation of the vestibulo-ocular reflex (VOR) gain with vergence. In *Proceedings of the 3rd Joint Symposium on Neural Computation, Institute of Neural Computation,* pp. 202–221. University of California, San Diego, and Center for Neuromorphic Systems Engineering, Caltech, Pasadena, CA.

———. 1987. A neurophysiological view of autism. In *Neurobiological issues in autism* (eds. E. Schopler and G.B. Mesibov), pp. 258–324. Plenum Press, New York, NY.

———. 1995. Infantile autism. Part 1: MR imaging abnormalities and their neurobehavioral correlates. *Int. Pediatr.* **10:** 141–154.

———. 1997. Brainstem, cerebellar and limbic neuroanatomical abnormalities in autism. *Curr. Opin. Neurobiol.* **7:** 269–278.

Courchesne, E., S.A. Hillyard, and R.Y. Courchesne. 1977. P3 waves to the discrimination of targets in homogenous stimulus sequences. *Psychophysiology* **14:** 590–597.

Courchesne, E., N.A. Akshoomoff, and J. Townsend. 1990. Recent advances in autism. *Curr. Opin. Pediatr.* **2:** 685–693.

Courchesne, E., J. Townsend, N. Akshoomoff, O. Saitoh, R. Yeung-Courchesne, A. Lincoln, H. James, R. Haas, L. Schreibman, and L. Lau. 1994. Impairment in shifting attention in autistic and cerebellar patients. *Behav. Neurosci.* **108:** 848–865.

Courchesne, E., N.A. Akshoomoff, J. Townsend, and O. Saitoh. 1995. A model system for the study of attention and the cerebellum: Infantile autism. Perspectives of event-related potentials research. *Electroencephalogr. Clin. Neurophysiol.* (Suppl. 44): 315–325.

Courtney, S.M., L.G. Ungerleider, K. Keil, and J.V. Haxby. 1996. Object and spatial visual working memory activate separate neural systems in human cortex. *Cereb. Cortex* **6:** 39–49.

Crispino, L. and T.H. Bullock. 1984. Cerebellum mediates modality-specific modulation of sensory responses of the midbrain and forebrain in rat. *Proc. Natl. Acad. Sci.* **81:** 2917–2920.

Darlot, C. 1993. The cerebellum as a predictor of neural messages—I. The stable estimator hypothesis. *Neuroscience* **56:** 617–646.

Decety, J., D. Perani, M. Jeannerod, V. Bettinardi, B. Tadary, R. Woods, J.C. Mazziotta, and F. Fazio. 1994. Mapping motor representations with positron emission tomography. *Nature* **371:** 600–602.

Deiber, M.-P., V. Ibañez, N. Sadato, and M. Hallett. 1996. Cerebral structures participating in motor preparation in humans: A positron emission tomography study. *J. Neurophysiol.* **75:** 233–247.

Dugas, C. and A.M. Smith. 1992. Responses of cerebellar Purkinje cells to slip of a hand-held object. *J. Neurophysiol.* **67:** 483–495.

Ellerman, J.M., D. Flament, S.-G. Kim, Q.-G. Fu, H. Merkle, T.J. Ebner, and K. Ugurbil. 1994. Spatial patterns of functional activation of the cerebellum investigated using high field (4T) MRI. *NMR Biomed.* **7:** 63–68.

Fiez, J.A., E.A. Raife, D.A. Balota, J.P. Schwarz, M.E. Raichle, and S.E. Petersen. 1996. A positron emission tomography study of the short-term maintenance of verbal information. *J. Neurosci.* **16:** 808–822.

Flament, D., J.M. Ellermann, S.-G. Kim, K. Ugurbil, and T.J. Ebner. 1996. Functional magnetic resonance imaging of cerebellar activation during the learning of a visuomotor dissociation task. *Hum. Brain Mapping* **4:** 210–226.

Fox, P.T., M.E. Raichle, and W.T. Thach. 1985. Functional mapping of the human cerebellum with positron emission tomography. *Proc. Natl. Acad. Sci.* **82:** 7462–7466.

Friston, K.J., C.D. Frith, R.E. Passingham, P.F. Liddle, and R.S.J. Frackowiak. 1992. Motor practice and neurophysiological adaptation in the cerebellum: A positron tomography study. *Proc. R. Soc. Lond. B. Biol. Sci.* **248:** 223–228.

Gao, J.-H., L.M. Parsons, J.M. Bower, J. Xiong, J. Li, and P.T. Fox. 1996. Cerebellum implicated in sensory acquisition and discrimination rather than motor control. *Science* **272:** 545–547.

Ghez, C. 1991. The cerebellum. In *Principles of neural sciences* 3rd ed. (ed. E.R. Kandel, J.H. Schwartz, and T.M. Jessell), pp. 626–646. Elsevier Science Publishing Co., New York, NY.

Glickstein, M., N. Gerrits, I. Kralj-Hans, B. Mercier, J. Stein, and J. Voogd. 1994. Visual pontocerebellar projections in the macaque. *J. Comp. Neurol.* **349:** 51–72.

Grafton, S.T., R.P. Woods, and M. Tyszka. 1994. Functional imaging of procedural motor learning: Relating cerebral blood flow with individual subject performance. *Hum. Brain Mapping* **1:** 221–234.

Grasby, P.M., C.D. Frith, K.J. Friston, C. Bench, R.S.J. Frackowiak, and R.J. Dolan. 1993. Functional mapping of brain areas implicated in auditory-verbal memory function. *Brain* **116:** 1–20.

Haines, D.E. and E. Dietrichs. 1987. On the organization of interconnections between the cerebellum and hypothalamus. In *New concepts in cerebellar neurobiology* (ed. J.E. King), pp. 113–149. Alan R. Liss, New York, NY.

Holmes, G. 1939. The cerebellum of man. *Brain* **7:** 121–172.

Ikai, Y., M. Takada, Y. Shinonaga, and N. Minzuno. 1992. Dopaminergic and non-dopaminergic neurons in the ventral tegmental area of the rat project, respectively, to the cerebellar cortex and deep cerebellar nuclei. *Neuroscience* **51:** 719–728.

Ito, M. 1984. *The cerebellum and neural control.* Raven Press, New York, NY.

Itoh, K. and N. Mizuno. 1979. A cerebello-pulvinar projection in the cat as visualized by the use of antero-grade transport of horseradish peroxidase. *Brain Res.* **171:** 131–134.

Ivry, R.B. and S.W. Keele. 1989. Timing functions of the cerebellum. *J. Cognit. Neurosci.* **1:** 136–152.

Jenkins, H.I., D.J. Brooks, P.D. Nixon, R.S.J. Frackowiak, and R.E. Passingham. 1994. Motor sequence learning: A study with positron emission tomography. *J. Neurosci.* **14:** 3775–3790.

Kanner, L. 1943. Autistic disturbances of affective contact. *Nervous Child* **2:** 217–250.

Kim, S.G., K. Ugurbil, and P.L. Strick. 1994. Activation of a cerebellar output nucleus during cognitive processing. *Science* **265:** 949–951.

King, J.E. 1987. *New concepts in cerebellar neurobiology.* Alan R. Liss, New York, NY.

Kitano, K., Y. Ishida, T. Ishikawa, and S. Murayama. 1976. Responses of extralemniscal thalamic neurones to stimulation of the fastigial nucleus and influences of the cerebral cortex in the cat. *Brain Res.* **106:** 172–175.

Klingberg, T., R. Kawashima, and P.E. Roland. 1996. Activation of multi-modal cortical areas underlies short-term memory. *Eur. J. Neurosci.* **8:** 1965–1971.

Krupa, D.J., J.K. Thompson, and R.F. Thompson. 1993. Localization of a memory trace in the mammalian brain. *Science* **260:** 989–991.

Liu, F.-Y., J.-T. Qiao, and N. Dafny. 1993. Cerebellar stimulation modulates thalamic noxious-evoked responses. *Brain Res. Bull.* **30:** 529–534.

Llinás, R. and C. Sotelo. 1992. *The cerebellum revisited.* Springer-Verlag, New York, NY.

Logan, C.G. and S.T. Grafton. 1995. Functional anatomy of human eyeblink conditioning determined with regional cerebral glucose metabolism and positron-emission tomography. *Proc. Natl. Acad. Sci.* **92:** 7500–7504.

Lynch, J.C., J.E. Hoover, and P.L. Strick. 1994. Input to the primate frontal eye field from the substantia nigra, superior colliculus, and dentate nucleus demonstrated by transneuronal transport. *Exp. Brain Res.* **100:** 181–186.

Martin, A., J.V. Haxby, F.M. LaLonde, C.O. Wiggs, and L.G. Ungerleider. 1995. Discrete cortical regions associated with knowledge of color and knowledge of action. *Science* **270:** 102–105.

McCormick, D.A. and R.T. Thompson. 1984. Cerebellum: Essential involvement in the classically conditioned eyelid response. *Science* **223:** 296–299.

Mellett, E., N. Tzourio, M. Denis, and B. Mazoyer. 1995. A positron emission tomography study of visual and mental spatial exploration. *J. Cognit. Neurosci.* **7:** 433–445.

Miall, R.C., D.J. Weir, D.M. Wolpert, and J.F. Stein. 1993. Is the cerebellum a Smith predictor? *J. Mot. Behav.* **25:** 203–216.

Middleton, F.A. and P.L. Strick. 1994. Anatomical evidence for cerebellar and basal ganglia involvement in higher cognitive function. *Science* **266:** 458–461.

Molchan, S.E., T. Sunderland, A.R. McIntosh, P. Herscovitch, and B.G. Schreurs. 1994. A functional anatomical study of associative learning in humans. *Proc. Natl. Acad. Sci.* **91:** 8122–8126.

Moruzzi, G. and H.W. Magoun. 1949. Brainstem reticular formation and activation of the EEG. *Electroencephalogr. Clin. Neurophysiol.* **1:** 455–473.

Moscovitch, M., S. Kapur, S. Köhler, and S. Houle. 1995. Distinct neural correlates of visual long-term memory for spatial location and object identity: A positron emission tomography study in humans. *Proc. Natl. Acad. Sci.* **92:** 3721–3725.

Nagahama, Y., H. Fukuyama, H. Yamauchi, S. Matsuzaki, J. Konishi, H. Shibasaki, and J. Kimura. 1996. Cerebral activation during performance of a card sorting test. *Brain* **119:** 1667–1675.

Nakai, M., C. Iadecola, D.A. Ruggiero, L.W. Tucker, and D.J. Reis. 1983. Electrical stimulation of cerebellar fastigial nucleus increases cerebral cortical blood flow without change in local metabolism: Evidence for an intrinsic system in the brain for primary vasodilation. *Brain Res.* **260:** 35–49.

Newman, P.P. and H. Reza. 1979. Functional relationships between the hippocampus and the cerebellum: An electrophysiological study of the cat. *J. Physiol. (Lond.)* **287:** 405–426.

Nichols, S., J. Townsend, and B. Wulfeck. 1995. Covert visual attention in language-impaired children. *Technical Report No. CND-9502,* Center for Research in Language, University of California at San Diego.

Nieuwenhuys, R., J. Voogd, and C. van Huijzen. 1988. *The human central nervous system.* Springer-Verlag, Berlin, Germany.

Parsons, L.M., P.T. Fox, J.H. Downs, T. Glass, T.B. Hirsch, C.C. Martin, P.A. Jerabek, and J.L. Lancaster. 1995. Use of implicit motor imagery for visual shape discrimination as revealed by PET. *Nature* **375:** 54–58.

Paulin, M.G. 1993. The role of the cerebellum in motor control and perception. *Brain Behav. Evol.* **41:** 39–50.

Petersen, S.E., P.T. Fox, M.I. Posner, M. Mintun, and M.E. Raichle. 1989. Positron emission tomographic studies of the processing of single words. *J. Cognit. Neurosci.* **1:** 153–170.

Petit, L., C. Orssaud, N. Tzourio, F. Crivello, A. Berthoz, and B. Mazoyer. 1996. Functional anatomy of a prelearned sequence of horizontal saccades in humans. *J. Neurosci.* **16:** 3714–3726.

Posner, M.I., J.A. Walker, F.J. Friedrich, and R.D. Rafal. 1984. Effects of parietal injury on covert orienting of attention. *J. Neurosci.* **4:** 1863–1874.

Raichle, M.E., J.A. Fiez, T.O. Videen, A.K. MacLeod, J.V. Pardo, P.T. Fox, and S.E. Petersen. 1994. Practice-related changes in human brain functional anatomy during nonmotor learning. *Cereb. Cortex* **4:** 8–26.

Saint-Cyr, J.A. and D.J. Woodward. 1980a. Activation of mossy and climbing fiber pathways to the cerebellar cortex by stimulation of the fornix in the rat. *Exp. Brain Res.* **40:** 1–12.

———. 1980b. A topographic analysis of limbic and somatic inputs to the cerebellar cortex in the rat. *Exp. Brain Res.* **40:** 13–22.

Sasaki, K., Y. Matsuda, S. Kawaguchi, and N. Mizuno. 1972. On the cerebello-thalamo-cerebral pathway for the parietal cortex. *Exp. Brain Res.* **16:** 89–103.

Sasaki, K., K. Jinnai, H. Gemba, S. Hashimoto, and N. Mizuno. 1979. Projection of the cerebellar dentate nucleus onto the frontal association cortex in monkeys. *Exp. Brain Res.* **37:** 193–198.

Schmahmann, J.D. 1996. From movement to thought: Anatomic substrates of the cerebellar contribution to cognitive processing. *Hum. Brain Mapping* **4:** 174–198.

———. 1997. Rediscovery of an early concept. In *International review of neurobiology. The cerebellum and cognition* (ed. J.D. Schmahmann), Vol. 41, Academic Press, San Diego, CA. (In press.)

Schmahmann, J.D. and D.N. Pandya. 1989. Anatomical investigation of projections to the basis pontis from the posterior parietal association cortices in rhesus monkey. *J. Comp. Neurol.* **289:** 53–73.

Seitz, R.J., A.G.M. Canavan, L. Yágüez, H. Herzog, L. Tellmann, U. Knorr, Y. Huang, and V. Hömberg. 1994. Successive roles of the cerebellum and premotor cortices in trajectorial learning. *NeuroReport* **5:** 2541–2544.

Siegel, P. and J.G. Wepsic. 1974. Alteration of nociception by stimulation of cerebellar structures in the monkey. *Physiol. Behav.* **13:** 189–194.

Snider, R.S. 1950. Recent contributions to the anatomy and physiology of the cerebellum. *Arch. Neurol. Psychiatry* **64:** 196–219.

———. 1967. Functional alterations of cerebral sensory areas by the cerebellum. In *Progress* in brain research. *The cerebellum* (ed. C.A. Fox and R.S. Snider), pp. 322–333, Vol. 25. Elsevier, Amsterdam, The Netherlands.

Snider, R.S. and J. Mitra. 1970. Cerebellar influences on units in sensory areas of cerebral cortex. In *The cerebellum in health and disease* (ed. W.S. Fields and W. D. Willis), pp. 319–331, 339. W.H. Green, St. Louis, MO.

Snyder, L.H., D.M. Lawrence, and W.M. King. 1992. Changes in vestibulo-ocular reflex (VOR) anticipate changes in vergence angle in monkey. *Vision Res.* **32:** 569–575.

Squires, K.C., S.A. Hillyard, and P.H. Lindsay. 1973. Vertex potentials evoked during auditory signal detection: Relation to decision criteria. *Percept. Psychophys.* **14:** 265–272.

Stuss, D.T., T. Shallice, M.P. Alexander, and T.W. Picton. 1995. A multidisciplinary approach to anterior attention functions. *Ann. N.Y. Acad. Sci.* **769:** 191–211.

Swanson, J.M., M.I. Posner, S. Potkin, S. Bonforte, D. Youpa, C. Fiore, D. Cantwell, and F. Crinella. 1991. Activating tasks for the study of visual-spatial attention in ADHD: A cognitive anatomic approach. *J. Child Neurol.* **6:** S119–S127.

Talairach, J. and P. Tournoux. 1988. *Co-planar stereotaxic atlas of the human brain.* Thieme Medical Publishers, New York, NY.

Thielert, C.-D. and P. Thier. 1993. Patterns of projections from the pontine nuclei and the nucleus reticularis tegmenti pontis to the posterior vermis in the rhesus monkey: A study using retrograde tracers. *J. Comp. Neurol.* **337:** 113–126.

Townsend, J., E. Courchesne, and B. Egaas. 1996a. Slowed orienting of covert visual-spatial attention in autism: Specific deficits associated with cerebellar and parietal abnormality. *Dev. Psychopathol.* **8:** 563–584.

Townsend, J., N. Singer, and E. Courchesne. 1996b. Visual attention abnormalities in autism: Delayed orienting to location. *J. Int. Neuropsychol. Soc.* **2:** 541–550.

Vilensky, J.A. and G.W. Van Hoesen. 1981. Corticopontine projections from the cingulate cortex in the rhesus monkey. *Brain Res.* **205:** 391–395.

Watson, P.J. 1978. Nonmotor functions of the cerebellum. *Psychol. Bull.* **85:** 944–967.

Welsh, J.P. and J.A. Harvey. 1992. The role of the cerebellum in voluntary and reflexive movements: History and current status. In *The cerebellum revisited* (ed. R. Llinás and C. Sotelo), pp. 301–334. Springer-Verlag, New York, NY.

Williams, R.W. and K. Herrup. 1988. The control of neuron number. *Annu. Rev. Neurosci.* **11:** 423–453.

RESEARCH

Impaired Capacity of Cerebellar Patients to Perceive and Learn Two-Dimensional Shapes Based on Kinesthetic Cues

Yuri Shimansky, Marian Saling, David A. Wunderlich, Vlastislav Bracha, George E. Stelmach, and James R. Bloedel[1]

Division of Neurobiology
Barrow Neurological Institute
Phoenix, Arizona
The Motor Control Laboratory
Arizona State University
Tempe, Arizona

Abstract

This study addresses the issue of the role of the cerebellum in the processing of sensory information by determining the capability of cerebellar patients to acquire and use kinesthetic cues received via the active or passive tracing of an irregular shape while blindfolded. Patients with cerebellar lesions and age-matched healthy controls were tested on four tasks: (1) learning to discriminate a *reference* shape from three others through the repeated tracing of the *reference* template; (2) reproducing the *reference* shape from memory by drawing blindfolded; (3) performing the same task with vision; and (4) visually recognizing the *reference* shape. The cues used to acquire and then to recognize the *reference* shape were generated under four conditions: (1) "active kinesthesia," in which cues were acquired by the blindfolded subject while actively tracing a *reference* template; (2) "passive kinesthesia," in which the tracing was performed while the hand was guided passively through the template; (3) "sequential vision," in which the shape was visualized by the serial exposure of small segments of its outline; and (4) "full vision," in which the entire shape was visualized. The sequential vision condition was employed to emulate the sequential way in which kinesthetic information is acquired while tracing the *reference* shape. The results demonstrate a substantial impairment of cerebellar patients in their capability to perceive two-dimensional irregular shapes based only on kinesthetic cues. There also is evidence that this deficit in part relates to a reduced capacity to integrate temporal sequences of sensory cues into a complete image useful for shape discrimination tasks or for reproducing the shape through drawing. Consequently, the cerebellum has an important role in this type of sensory information processing even when it is not directly associated with the execution of movements.

Introduction

A century of investigations examining the characteristics of cerebellar systems and the deficits produced by cerebellar lesions has provided many insights into cerebellar function (Dow and Moruzzi 1958; Gilman et al. 1981), including the fact that this structure is involved in processing information derived from representing a wide variety of sensory modalities via an involvement of many central projections (Bloedel and Courville 1981). These inputs include projections arising from muscle spindles, tendon organs, and joint receptors, as well as from a wide variety of cutaneous mechanoreceptors (Bloedel and Courville 1981; Granit 1970). Most discussions regarding the pro-

[1]Corresponding author.

cessing of these inputs have emphasized their role in the regulation of movements (Gilman et al. 1976, 1981; Granit 1970; MacKay and Murphy 1979; Kawato and Gomi 1992; Stein and Glickstein 1992).

The importance of the sensory processing produced by the cerebellum was emphasized recently by Paulin (1993), who suggested that this cerebellar-dependent integration could contribute substantially to the ongoing control of movement. Perhaps because the processing of sensory information by components of the motor system is so tightly coupled to the generation and control of movements, few investigations have been undertaken to examine the contributions of these central structures to the perceptual phenomena resulting from the activation of these same afferent inputs. However, the divergent, parallel organization of cerebellar efferent systems to many central sites that are not related directly to movement generation (see Haines 1997) suggests that the processing of sensory information by this structure does not necessarily have to be associated with a motor action. Recently, Gao et al. (1996) published a provocative study in which they provided evidence that the cerebellum could contribute to the perception of cutaneous stimuli even if movement or the intent or need to move was excluded as a factor. This result suggests that, independent of the capacity of a central nervous system component to employ this sensory processing in motor control, it also may contribute to sensory perception. In addition, there is progressively mounting evidence that the cerebellum may have a role in the recognition and/or perception of different stimuli even when these stimuli are dissociated from the need to generate movements. For example, Canavan et al. (1994) demonstrated that cerebellar patients had a decreased capacity to recognize abstract visual cues when compared with normal subjects. The experiment reported here examines whether the cerebellum contributes to the perception of kinesthetic cues.

It is well known that kinesthetic cues are employed during the execution of normal movements (Cordo and Flanders 1989; Lovelace 1989; Tillery et al. 1991; Soechting and Flanders 1992). These cues are processed rapidly enough to provide highly relevant information required for the on-line control of volitional limb movements (Cordo 1988, 1990; Flanders and Cordo 1989). It is also clear that coding of target and limb position can employ coordinate systems that are based on kinesthetic information (Tillery et al. 1991; Darling and Miller 1993). Many of these applications of kinesthetic information require transformations of sensory inputs from the periphery into reference frames that are suitable not only for generating movements but also for generating visual representations of body position or the characteristics of external objects or conditions being evaluated.

The importance of the cerebellum for the transformation of sensory cues into a representation within a coordinate system appropriate for controlling movements has been emphasized extensively by Pellionisz (1985), Pellionisz and Llinas (1985), and Paulin (1993). Roll and Gilhodes (1995) showed recently that important transformations may occur specifically related to the processing of kinesthetic information. They demonstrated that transformations of proprioceptive cues may be responsible for representing spatial features of limb trajectories in joint space coordinates. Furthermore, kinesthetic information is likely required for deriving optimally an internal representation of the position of visualized targets in extrapersonal space (Tillery et al. 1991). Given the substantial amount of kinesthetic information received by the cerebellum and its capacity to affect interactions in motor pathways, this structure could contribute to these functionally critical transformations. If so, and if the premise of our experiment is correct, this cerebellar-dependent processing also could contribute to the perception of kinesthetic cues.

One of the few previous studies addressing the issue of the participation of the cerebellum in kinesthesia was performed by Grill et al. (1994). These investigators examined the perceived duration, amplitude, and velocity of passively imposed finger movements. Statistically significant deficits in the capacity to differentiate between joint displacements of different durations and between different velocities of ramp displacements were reported. Although these deficits were not profound, they suggested the possibility that the cerebellum may in fact contribute to the perception of kinesthetic cues. The present study pursued the role of the cerebellum in kinesthesia from a somewhat different perspective. Rather than examining the perception of relatively simple cues, our paradigm requires that a higher order of information processing be performed—the formation of an internal representation of the shape by integrating temporal sequences of kinesthetic cues received through the active or passive tracing of an irregular shape while blindfolded. We hypothesized that the optimal on-line processing of temporal and spatial kinesthetic cues required for the execution of this task involves integration performed by the cerebellum. The experiment employed a learning paradigm to determine whether repeated exposure to the same shape improved the capacity of the subjects to recognize it among other dissimilar shapes, a process likely to depend on the establishment of an internal representation of the original image.

In the first task of the experiment, we compared the capacity of normal subjects and cerebellar patients to improve their discrimination of a *reference* template from three others using one of four different ways to assess its shape and compare it with the test templates: active tracing of the *reference* template with their eyes closed; passive tracing of the *reference* with their eyes closed; sequential visualization of successive segments of the template; and complete visualization of the *reference* template. This design made it possible to compare the acquisition and evaluation of cues using kinesthetic information with those observed when visual information was employed. Second, to determine whether subjects could draw an outline of the *reference* shape once acquired under each of the above conditions and to test whether visualization of the hand during drawing made a difference in the performance of the subjects, we compared their capacity to reproduce the *reference* shape by drawing, first while blindfolded and then with vision. In the last task, the capacity to recognize the *reference* shape visually was tested. This part of the experiment evaluated whether the subject could recognize the *reference* shape when no kinesthetic cues were required. The present study is the first to assess the performance of these types of complex tasks requiring kinesthetic cues in patients with cerebellar disorders.

The data will demonstrate that unlike normal subjects, cerebellar patients are substantially impaired in their capability to perceive two-dimensional irregular shapes based only on kinesthetic cues. This impairment is manifested in all the above tasks. Evidence also will be presented indicating that this deficit in part relates to a reduced capacity to integrate temporal sequences of sensory cues into a complete image useful for performing recognition tasks or for representing the shape through drawing. Preliminary data from this study has been presented previously (Shimansky et al. 1996).

Materials and Methods

Experiments were performed with seven patients having chronic, isolated cerebellar lesions and seven age-matched healthy controls, all of whom participated after providing full consent on the basis of all guidelines specified by the Institutional Review Board for Human Research. A brief summary of their clinical characteristics is given in Table 1. As outlined in Figure 1, the experiment was performed in two phases, the shape acquisition/discrimination phase and the drawing and visual recognition phase. Throughout both phases of the experiment, the subject was seated comfortably in front of a table on which the experimental manipulations were performed.

SHAPE ACQUISITION/DISCRIMINATION PHASE

In the first phase of the experiment, subjects were tested on their capacity to recognize an irregularly shaped object based on information regarding its shape acquired under four different conditions. In the first two conditions, blindfolded subjects acquired information about the shape of an object either by moving a pen actively through a grooved template three times [active kinesthesia (AK)] or by holding a pen-shaped manipulandum that was moved passively through the grooved template while the elbow was suspended above the surface of the table [passive kinesthesia (PK)]. In the third condition, the subject was exposed to the object's shape using a sequential vision (SV) paradigm. For sequential presentation of the object's shape, the template was covered by a nontransparent disk with a 10-mm gap exposing only ~3% of the outline of the template at any one time. The disk was then rotated at a speed of 4 rpm. (one full turn in 15 sec), a speed that matched that of the active or passive movements performed during the first two conditions. In the fourth condition, the shape of the object was visualized in its entirety [full vision (FV)].

A different set of template shapes was used in each of these acquisition/discrimination paradigms. The degree of template shape complexity was the same in AK, PK, and SV conditions,

Table 1: Clinical characteristics of patients in this study

Subject	Diagnosis	Radiological findings[a]	Deficits
1. G.M.: male, 49 years	right cerebellar arteriovenous malformation (resected in 1992)	CT: large lesion involving the right cerebellar hemispheric cortex, vermis, and the cerebellar nuclei	dysarthria, ataxia of right arm and leg, a significant intentional tremor, and a slight ptosis (1 mm) on the right side
2. C.R.: male, 58 years	left cerebellar cortex tumor (tumor and undersurface of the tentorium excised in 1988)	MRI: large tumor (40 × 30 mm) with slight involvement of the left nuclei	slight gait ataxia and mild ataxia and target oscillations with his left arm
3. R.F.: male, 44 years	left hemisphere dysplastic gangliocytoma (resected in 1991)	MRI: dysplastic gangliocytoma (30 × 70 mm) involving the left hemisphere, midline, and a small part of the right hemisphere	slightly ataxic gait and upper extremity ataxia prior to surgery; presently does not display signs of cerebellar deficits
4. T.J.: male, 53 years	right cerebellar cortical lesion (surgically excised in 1995)	MRI: 11-mm lesion in the dorsolateral aspect of the right cerebellar hemisphere	upper extremity cerebellar ataxia, as well as an intentional tremor in the right arm
5. H.J.: female, 48 years	posterior fossa pilocystic astrocytoma (at age of 13)	CT: only rudimentary parts of the cerebellum remain	severe lower and upper limb ataxia with intentional tremor in the right arm
6. E.A.: female, 45 years	left cerebellar recurrent tumor (completely excised in 1995)	MRI: large lesion involving left cerebellar hemisphere and large parts of the vermis	slight gait ataxia and mild ataxia of the upper extremities
7. D.T.: female, 53 years	right cerebellar tumor (excised in 1996)	MRI: large tumor with slight involvement of the right cerebellar nuclei	extremely mild gait ataxia and slight ataxia of the right arm; general weakness of right arm

[a](CT) Computerized tomography; (MRI) magnetic resonance imaging.

and it was slightly higher (greater number of protrusions) in the FV condition. All images used in this study were unfamiliar to the subjects before the experiment. The time interval between experiments involving the performance of the different paradigms by the same subject was never less than a week.

For each of the four conditions tested, a series of four sets of trials was performed. The design of a typical set is characterized in Figure 2. At the beginning of each set, a subject was asked to scan a *reference* profile (trace it three times in AK and PK conditions, visually examine its shape during three full turns of the disk in SV condition, and visualize the shape for 30 sec in FV condition) and memorize its shape. Then a sequence of five *test* profiles was presented in pseudorandom order. Two of the five were the same as the *reference* profile and three were different. The subjects were required to scan each *test* profile (always in the same way as the *reference* profile) and indicate whether its shape was identical to the *reference* one. Then, the number of errors made in the discrimination procedure for the set was recorded.

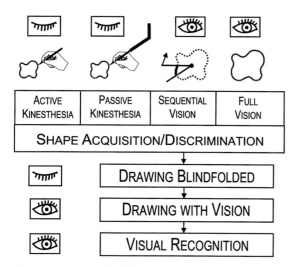

Figure 1: Protocol of the experiment. The pictures denote four alternative conditions under which the first task (acquisition/discrimination) was performed (see Materials and Methods). Conditions for the other three tasks (indicated by the pictures on the *left*) were the same independent of the condition of the first task.

This procedure was repeated three additional times, providing a succession of four sets of trials over which any change in the number of errors was assessed.

DRAWING AND VISUAL RECOGNITION PHASE

Before the drawing tasks began, the subject was asked to scan the *reference* shape to which he/she was exposed previously at the initiation of the acquisition/discrimination phase. This was done for each of the four conditions described above (AK, PK, SV, FV). The subject then was asked to draw it from memory five consecutive times while blindfolded. The second drawing task was identical to the first with the exception that the subject drew the profile using vision. In the final test of this phase, each subject was required to recognize visually the shape of the *reference* profile without any additional exposure to it. In the first trial, pictures of the same four shapes (including the *reference*) were presented serially. If the subject identified the *reference* shape correctly in the first trial, the task was considered to be completed, and no additional trials were performed. If the answer was not correct, the incorrectly-identified shape was removed from the group, and another trial was performed. If no shape was identified as the *reference*, the last of the four shapes presented, which was never the *reference*, was removed. In the second trial, the three remaining shapes were shown together. If again a mistake was made, the misidentified shape was removed, and a third trial was performed in which the subject chose between the remaining two shapes. The subject's score on this recognition task was the number of trials required to identify the *reference* shape. When an error was made on the third trial, the subject received the score of 4.

DATA ANALYSIS

For each of the four conditions in the discrimination task (AK, PK, SV, and FV), statistical comparison of the numbers of errors (failure to recognize the *reference* profile or identification of a non-reference profile as the *reference*) made in each successive set was performed using the standard

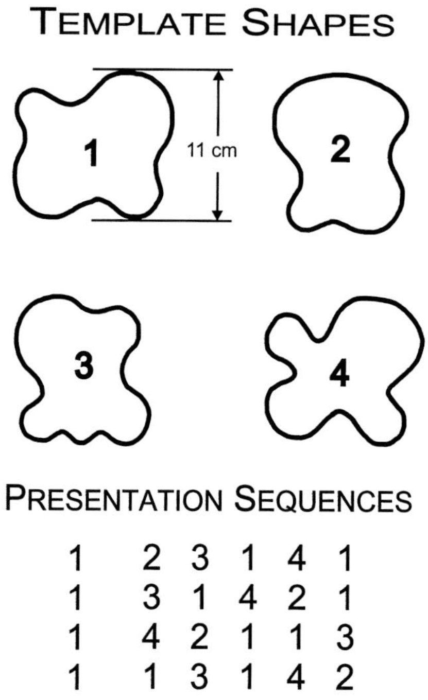

TEMPLATE SHAPES

PRESENTATION SEQUENCES

1	2	3	1	4	1
1	3	1	4	2	1
1	4	2	1	1	3
1	1	3	1	4	2

Figure 2: An example of a set of template shapes, this one used in the PK condition. Shape 1 was the *reference* shape, and shapes 2–4 were *test* shapes. The four sequences of shape presentations used in each of four successive sets of trials in the shape acquisition/discrimination task are shown below. Each sequence began with the subject tracing the reference shape. This was followed by a succession of five trials in which the shapes were presented in a pseudorandom order. See Materials and Methods for explanation.

t-test (CSS Statistica) to compare the performance of cerebellar patients and control subjects, with $P < 0.05$ selected as the basis for determining significance. These groups were compared based on their errors in the first set of trials, the last set, and the total errors across all sets. Also, the number of errors in the last set was compared with that in the first set within each group of subjects to estimate the effect of any learning. In addition to the intergroup comparisons, the performance of each group was assessed relative to the chance level. The latter was calculated as the mathematical expectation of the number of errors produced if the identification of the subjects of each shape (in a set of two *reference* and three *test* shapes) as "reference" or "nonreference" were totally random (the independent probability of either choice being equal to 0.5). The mathematical expectation of the number of errors in one set of trials and the total across all four sets is equal to 2.5 and 4, respectively, and the corresponding standard deviations are approximately equal to 1.12 and 2.24, respectively.

For the drawing and visual recognition phase, the resemblance of each subject's drawings to the corresponding *reference* shape was compared qualitatively between the group of cerebellar patients and the group of control subjects. The differences relevant to this manuscript were substantial enough so that inferences could be drawn easily from the qualitative differences in the performance between the two groups of subjects. Because no scoring of these qualitative features was performed, the evaluator was not blind relative to the nature of the subject. For the visual recognition test, the number of trials required to recognize the *reference* shape under each condition was compared between cerebellar patients and control subjects based on the standard *t*-test.

Results

SHAPE ACQUISITION/DISCRIMINATION TASK

In the first series of experiments performed in the four groups of cerebellar patients and age- and sex-matched control subjects, each individual was required to perform four sets of trials. In each trial the subjects scanned the *reference* template and then performed a discrimination task with a total of five templates with four different shapes that were presented in a pseudorandom order. This was done for all four acquisition/discrimination conditions (see Fig. 1). As described more thoroughly in Materials and Methods, the capacity of each subject to discriminate between the *reference* shape and three other nonreference shapes was evaluated over four consecutive sets of trials to assess whether the number of errors (failure to recognize the *reference* shape or identification of a nonreference shape as the *reference*) decreased as this task was practiced. The results of this part of the experiment are shown in Figure 3 and summarized in Tables 2–5.

The data obtained when shape discrimination was performed via active kinesthesia (Fig. 3A) and passive kinesthesia (Fig. 3B) were very similar. In both conditions, the numbers of errors made in the first set of trials were not significantly different ($P > 0.1$, *t*-test, also used in subsequent comparisons) between cerebellar patients and normal subjects or from the chance level. However, in contrast to control subjects, whose number of errors in the last (fourth) set of trials was significantly different from chance level and less than in the first set of trials ($P < 0.01$), the cerebellar patients were unable to significantly decrease the number of errors over the four sets of trials ($P > 0.1$). The total number of errors made by the cerebellar patients was significantly higher than that made by the control subjects ($P < 0.01$). When the subjects were

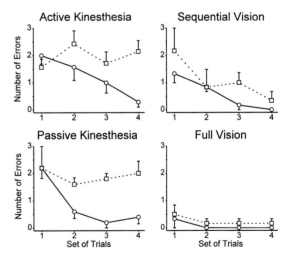

Figure 3: Learning curves for the shape discrimination task. The number of errors (failure to recognize the *reference* shape or the designation of a *test* shape as the *reference*) made by cerebellar patients (□) and control subjects (○) are plotted against the number of the set of trials. The experimental condition is shown above each graph. Error bars indicate the s.e.

Table 2: Control subjects vs. cerebellar patients

Condition	Set 1 errors	Set 2 errors	Set 3 errors	Set 4 errors	Total errors	Recognition errors
Active kinesthesia	$P \leq 0.34$	$P \leq 0.24$	$P \leq 0.24$	$P \leq 0.01$	$P \leq 0.02$	$P \leq 0.01$
Passive kinesthesia	$P \leq 1.00$	$P \leq 0.02$	$P \leq 0.01$	$P \leq 0.02$	$P \leq 0.01$	$P \leq 0.01$
Sequential vision	$P \leq 0.38$	$P \leq 1.00$	$P \leq 0.06$	$P \leq 0.34$	$P \leq 0.19$	$P \leq 0.02$
Full vision	$P \leq 0.74$	$P \leq 0.35$	$P \leq 0.35$	$P \leq 0.34$	$P \leq 0.31$	$P \leq 0.15$

asked after completion of the kinesthesia paradigms whether they had a mental visual image of the *reference* shape while performing the shape discrimination task, most control subjects and cerebellar patients gave a positive answer.

When shape acquisition/discrimination was performed in the sequential vision paradigm (Fig. 3C), neither the number of errors in the first or last set of trials, nor the total number of errors made by cerebellar patients was found to be significantly different from those made by control subjects ($P > 0.1$). The nonparametric Mann-Whitney U-test was used for these inter-group comparisons instead of the *t*-test because the standard deviation in the group of cerebellar patients was considerably larger than that for control subjects (e.g., 3.44 and 0.82, respectively, for total number of errors, $P < 0.01$, F test). This difference in standard deviation suggests that the group of cerebellar patients was not homogeneous with regard to their performance in the SV paradigm (see Discussion). The most important finding in this condition was that cerebellar patients, like the control subjects, improved their performance considerably over the four sets. Their number of errors was at the chance level ($P > 0.1$) in the first set of trials but was significantly different from the chance level ($P < 0.01$) in the last set of trials. However, the difference between the number of errors in the first and the last set of trials was found to be statistically significant for control subjects ($P < 0.01$) but only close to significant for cerebellar patients ($P = 0.07$), apparently attributable to the above-mentioned high variability of their results.

When shape acquisition was performed using full vision (Fig. 3D), both cerebellar patients and normal subjects had comparably low error rates from the very first set of trials. The performance of both groups was significantly different from chance (Tables 3 and 4). Because of this, rates of improvement were difficult to substantiate in this paradigm. Nevertheless, both groups showed a trend toward a decreased number of errors. The differences between the scores of cerebellar patients and control subjects were not significant and were noticeably smaller than in all other paradigms.

The data in this part of the study demonstrate that cerebellar patients differed from control subjects in their decreased capacity to improve their score in shape discrimination in both the AK and PK paradigms. However, cerebellar patients were able to improve in both the SV and FV paradigms. Last, in all paradigms but FV, the number of errors was consistently higher in cerebellar patients than

Table 3: Cerebellar subjects vs. chance level

Condition	Set 1 errors	Set 4 errors	Total errors
Active kinesthesia	$P \leq 0.10$	$P \leq 0.55$	$P \leq 0.08$
Passive kinesthesia	$P \leq 0.76$	$P \leq 0.48$	$P \leq 0.12$
Sequential vision	$P \leq 0.74$	$P \leq 0.01$	$P \leq 0.01$
Full vision	$P \leq 0.01$	$P \leq 0.01$	$P \leq 0.01$

Table 4: Control subjects vs. chance level

Condition	Set 1 errors	Set 4 errors	Total errors
Active kinesthesia	$P \leq 0.36$	$P \leq 0.01$	$P \leq 0.01$
Passive kinesthesia	$P \leq 0.65$	$P \leq 0.01$	$P \leq 0.01$
Sequential vision	$P \leq 0.07$	$P \leq 0.01$	$P \leq 0.01$
Full vision	$P \leq 0.01$	$P \leq 0.01$	$P \leq 0.01$

Table 5: *Comparison of set 1 vs. set 4*

Condition	Controls	Cerebellar patients
Active kinesthesia	$P \leq 0.01$	$P \leq 0.28$
Passive kinesthesia	$P \leq 0.01$	$P \leq 0.84$
Sequential vision	$P \leq 0.01$	$P \leq 0.07$
Full vision	$P \leq 0.35$	$P \leq 0.41$

normal subjects, even when improvement occurred.

DRAWING THE REFERENCE SHAPE

To assess further the characteristics of any internal representation of the *reference* shape developed during the acquisition/discrimination phase of the experiment, subjects in both groups were required to draw the *reference* shape first while blindfolded and then with their eyes open after acquiring it during the first part of the experiment. The most revealing observations on these qualitative data were the findings in the AK and FV groups. Figure 4 illustrates the drawings performed by control subjects and cerebellar patients when the template was acquired in the AK paradigm. Notice that control subjects were able to generate

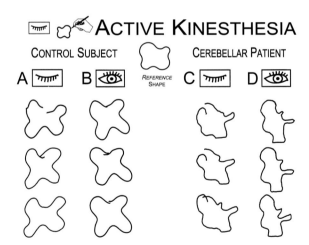

Figure 4: Examples of shapes drawn under the AK condition. (*A,B*) Examples of a control subject's drawings made while blindfolded and with vision, respectively. (*C,D*) Comparable drawings by a cerebellar patient. The *reference* shape (reduced in relative size) is shown above. The condition related to each diagram is the same as in Fig. 1.

Figure 5: Examples of shapes drawn under the FV condition. The layout and all designations are the same as in Fig. 4.

fairly reasonable representations of the template while blindfolded (Fig. 4A) as well as with eyes open (Fig. 4B). Although most subjects were able to draw a shape with the relative positions and orientation of the protrusions on the surface of the template in a relatively correct manner, their capacity to illustrate differences in the size of the various protrusions of the template was definitely better with the eyes open. Cerebellar patients had considerably more difficulty with this task. Consistent with their relatively high error rate on the shape discrimination task presented above, these subjects were unable to draw a reasonable representation of the template either while blindfolded (Fig. 4C) or with eyes open (Fig. 4D). However, it was apparent that their drawings show a noticeable trial-to-trial consistency.

When shape acquisition was performed with FV (Fig. 5), control subjects were able to draw a reasonable representation of the template either with eyes closed or open. In general the drawings with eyes closed included the different protrusions of the pattern in a reasonably accurate fashion (Fig. 5A). Cerebellar patients still displayed deficiencies in drawing an accurate representation of the template despite the fact that shape recognition was done with full vision. The drawings of the cerebellar patients while blindfolded usually consisted of a distorted representation of the template (Fig. 5C). Interestingly, this representation was qualitatively better than the representation of the template drawn after the shape was acquired in the AK paradigm (see Fig. 4C for comparison). However, even when the shape was drawn with full vision (Fig. 5D), the drawing was not accurately represented.

In this example, the patient drew an extra protrusion on the lobulated object.

VISUAL RECOGNITION OF THE REFERENCE SHAPE

After all subjects completed the drawing tasks with eyes open and blindfolded, they were asked to visually recognize the *reference* shape, as described in Materials and Methods. It should be noted that this recognition task was performed without re-scanning the *reference* template, as was the case in the first component of these experiments. In three of the four conditions (PK, SV, and FV), all cerebellar patients as well as controls were able to recognize the template visually after three presentations of the set of templates. In the AK paradigm, four of seven cerebellar patients were unable to recognize the shape of the *reference* template even in the third trial (when they had to choose between only two pictures). In both AK and PK paradigms, some cerebellar patients stated they were very surprised when seeing pictures of the actual template shapes because they were so different from the shapes they expected. The average number of trials required for visual recognition of the *reference* template is shown in Figure 6 for both cerebellar patients and controls (for those cerebellar patients who did not recognize the *reference* shape in three trials, the number of trials required for recognition was scored as 4, as described in Materials and Methods). In each paradigm except for FV, the cerebellar patients required a significantly greater number of recognition trials than the control subjects ($P < 0.02$, t-test). Note that the cerebellar patients revealed an impaired capacity to recognize the *reference* shape even when the template was acquired by sequential vision, despite the fact that they were capable of improving their rate of shape discrimination with practice in the first phase of the experiment.

Discussion

SUMMARY OF RESULTS

AK AND PK PARADIGMS

First, in both of these conditions, the number of errors made by cerebellar patients in the first set of shape discrimination trials and that made by control subjects were not significantly different either from the chance level or from each other. Second, unlike control subjects, cerebellar patients did not significantly decrease the number of errors in shape discrimination as they practiced the *reference* template over four consecutive sets of trials, the number of errors in the last set still being at the chance level. Third, the capability of cerebellar patients to draw a *reference* shape or to recognize it visually when it was acquired under either of the kinesthesia paradigms was considerably worse than that of control subjects. Fourth, the capacity of cerebellar patients either to draw or to recognize visually a *kinesthetically* presented reference shape under either kinesthesia condition was considerably less than when the reference shape was presented *visually* (in the FV condition).

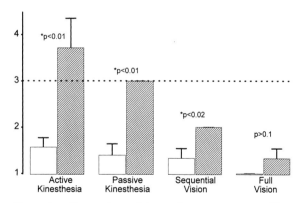

Figure 6: Comparison of the performance generated by cerebellar patients and control subjects on the visual recognition task. The bar graph illustrates the mean numbers of trials required for cerebellar patients and control subjects to recognize the *reference* shape acquired under the four different conditions (shown below the graph). The probability of the null hypothesis being correct based on the standard t-test is shown above the corresponding pair of bars (statistical significance is indicated by an asterisk). Error bars, 1 S.E.

FV AND SV PARADIGMS

First, in the shape discrimination phase, the number of errors made by *both* groups of subjects was significantly below the chance level by the end of the four series of trials in both vision paradigms. Second, in the FV paradigm, both groups of subjects could perform almost equally well over all four sets of trials with a low number of shape discrimination errors. In contrast, in the discrimination phase of the SV paradigm, the cerebellar performance of the patients was noticeably worse than that of the control subjects, and they also required significantly more trials to visually recognize the *reference* shape (in the fourth task). In the drawing tasks, the cerebellar patients performed notably better than they did in either of the kinesthesia paradigms. However, they remained qualitatively poorer in representing the *reference* shape than the normal subjects, particularly while blindfolded.

THE IMPAIRMENT IN THE ACQUISITION OF KINESTHETIC CUES

The results of this study, consistent with previous observations (Grill et al. 1994; Gao et al. 1996), unequivocally show that the processing of kinesthetic information is considerably impaired in cerebellar patients in this perceptual task. It should be emphasized that this deficit was present even though this information was not required for the execution of any movement or even the generation of any motor commands. An argument could be made that in the AK paradigm there may have been a relationship between the perceived kinesthetic cues and the generation of the tracing movement. However, if this occurred, it likely did not contribute to the results, as the performance of the cerebellar patients and normal subjects were comparable under both the AK and PK paradigms. Furthermore, the nature of this movement and its relation to the sensory cues is considerably different than under circumstances in which kinesthetic cues are required on-line to guide an ongoing movement. The fact that the performance of the subjects was the same under active and passive conditions also indicates that the deficit in shape discrimination was not related only to the generation of kinesthetic cues produced by the active movement required to trace the template. Interestingly, the learning rate appeared higher under the passive condition for normal subjects. This may be attributable to an interference between the generation of motor commands and the processing required for movement perception (Bairstow and Laszlo 1980). However, the overall similarity of the results obtained in both kinesthesia conditions (Fig. 3A, B) shows that the above interference was not critical for the perception of kinesthetic cues, at least in control subjects.

Although these experiments were not designed to differentiate between several possible reasons for the deficit observed in cerebellar patients, we propose that their inability to adequately reproduce the *reference* shape by drawing either blindfolded or with their eyes open after acquiring the shape via kinesthetic cues reflects an impairment in the processing of this type of information rather than an abnormality of some other higher order function. For example, this deficit could be explained by an impairment of retaining the image in memory, recalling it, and/or actually executing the drawing itself (attributable to the motor disorder). These possibilities are unlikely, however, given the fact that the patients could do reasonably well under the FV condition. Furthermore, it was shown in one of our recent experiments (Timmann, et al. 1996) that patients with cerebellar pathology not only could perform a mental rotation of an image and draw it in the rotated position after acquiring it by tracing with full vision, but they also could improve their performance of this task over successive trials.

Therefore, although other processes may be affected, we favor the view that the primary problem affecting the performance of the patients relates to an impairment in establishing a reliable internal representation of the *reference* shape. Interestingly, the deficit in these experiments is much greater than shown in the studies of Grill et al. (1994). They demonstrated a somewhat reduced capability to discriminate the duration and velocity of two successive passive finger movements. However, no significant deficit in the discrimination of movement amplitude was found. These smaller deficits in the discrimination of kinesthetic cues may be related to the difference between the complexity of the tasks in their experiment and that of our paradigms. In the studies of Grill et al. (1994), it was not necessary for the subjects to formulate the shape of an object to be used in subsequent parts of the experiment. This fact, together with the impairment observed in our study, suggests that the most significant deficit is related to a higher level of kinesthetic information processing than required only for discrimination of movement parameters. As stated above, we propose that this deficit is related to the generation of the internal representation of the image.

The possibility that the required transformations necessary for establishing this representation may occur but may be impaired is suggested from the drawing and recognition phase of the experiment. As shown in Figure 4, the patients acquiring the shape in the AK paradigm did not draw random representations of the template. Rather, certain features of the template were drawn, and the successive drawings were consistent. This would not be expected if the subject were merely guessing. It is more likely that these drawings represent an internal image that was developed inadequately and inaccurately.

Our data do not address the important issue of functional localization in the cerebellum, because all of the patients displayed the deficits reported above. Consequently, it was impossible to make inferences by comparing observations in patient groups having different localization of their pathology. Furthermore, the variability of their performance in different components of the experiment was not substantial enough to permit any conclusions regarding localization, particularly given the size of the patient group. In this experiment the greatest emphasis was on employing patients that had substantial pathology limited to the cerebellum. Additional experiments with patients having a greater distribution of focal cerebellar pathology would be required to make any well-founded arguments regarding this issue.

TRANSLATION BETWEEN KINESTHETIC AND VISUAL INFORMATION

An important question regarding the creation of an internal representation of a template shape is whether the representation acquired in the kinesthesia paradigms necessarily should be established as a visual image. Such translation was required only for the visual recognition of a kinesthetically acquired *reference* shape. Because in the PK paradigm all cerebellar patients recognized the *reference* shape only after three trials independent of their score in the shape discrimination task, it is quite feasible that the translation from the kinesthetic to the visual modality was impaired. It is also pertinent that the ability of the patients to draw the *reference* shape without vision after having acquired it in either visual paradigm was considerably better than in the kinesthesia paradigms. Because this finding demonstrates that the transformation required to draw the shape on the basis of a visual image was not impaired substantially in cerebellar patients, their deficit in the drawing task after acquiring the shape in a kinesthesia paradigm reflected a failure to formulate a visual image, presumably attributable to the inadequacy of the transformations of kinesthetic cues required in this process. It should be noted that shape recognition based on kinesthetic cues need not require a translation into the visual domain on theoretical grounds. However, it is possible that the nervous system selects such a translation because it is accustomed to operating in this domain when assessing or learning shapes of objects. Neither the cerebellar patients nor the control subjects were given any specific instruction pertinent to this question. When asked after the experiment whether they created a visual image of the template while kinesthetically acquiring the shape, most subjects usually gave a positive answer. Consequently, it is likely that a kinesthetic–visual transformation is performed during shape acquisition in the discrimination phase, and also that the success in the recognition task depends on the capacity to form an image in the visual domain in most subjects. This interesting question regarding the role of a kinesthetic–visual transformation in the acquisition and discrimination of spatial images requires further investigation.

INTEGRATION OF SEQUENTIALLY RECEIVED INFORMATION INTO A COMPLETE IMAGE

In contrast with visual information, kinesthetic information about a spatial shape is always received in a sequential manner. Because of the anticipated deficits in the kinesthesia paradigms, the SV paradigm was employed to obtain insights as to whether deficits in the processing of kinesthetic cues were related to the fact that the information was presented serially. The fact that cerebellar patients made significantly more errors than control subjects in the visual recognition task in the SV but not in the FV paradigm strongly suggests the existence of a deficit in their capability to integrate temporal sequences of visuo-spatial cues. Such a deficit could also account for the abnormalities in the visual perception of motion found in patients with midline cerebellar lesions (Nawrot and Rizzo 1995). Our observation, together with cerebellar impairment of the patients in the kinesthesia paradigms, suggests that the cerebellum can contribute to the development of an internal representation through the temporal integration of sensory cues. Clearly, the integrity of this functional mechanism may be important not only for the perception of shapes via kinesthetic cues but also for the control and learning of movements sequences.

THE ISSUE OF LEARNING

One of the intriguing aspects of the deficit in the perception of kinesthetic cues examined in this study is that its manifestation was clearest when patients were required to improve their capacity to perform shape discrimination over several trials. Because normal subjects improved their performance with practice dramatically on the kinesthesia paradigms, and because their performance in the initial set on the FV paradigm was substantially better than in the first set of the kinesthesia conditions, the establishment of a usable internal representation of the *reference* shape based on kinesthetic cues likely requires practice. In contrast, shape perception based on a single exposure to an object is based on the integration of a temporal sequence of kinesthetic cues. We define the improvement in the ability to recognize the shape of an object based on this type of information through practice as *kinesthetic learning*, a process in which an internal representation of the shape of the object is established by integrating sequences of data acquired over repeated exposures to the object through the acquisition of kinesthetic cues. The findings that both cerebellar patients and control subjects had similar levels of performance after the first exposure to the *reference* shape and that only the former improved after subsequent exposures to it, suggests that the cerebellum may not only be critical for establishing internal representation of objects based on kinesthetic information, but also that this structure may contribute to the process of kinesthetic learning.

Acknowledgments

This work was supported by National Institutes of Health grants R01NS21958 and P01NS30013 and Flinn Foundation Grant and Biomedical Enrichment Grant 019-109-103-92.

The publication costs of this article were defrayed in part by payment of page charges. This article must therefore be hereby marked ''advertisement'' in accordance with 18 USC section 1734 solely to indicate this fact.

References

Bairstow, P.J. and J.I. Laszlo. 1980. Motor commands and the perception of movement patterns. *J. Exp. Psychol.: Hum. Percept. & Perform.* **6:** 1–12.

Bloedel, J.R. and J. Courville. 1981. A review of cerebellar afferent systems. In *Handbook of physiology, vol. II. Motor control* (ed. V.B. Brooks), pp. 735–830. Williams and Wilkins, Baltimore, MD.

Canavan, A.G.M., R. Sprengelmeyer, H.-C. Diener, and V. Homberg. 1994. Conditional associative learning is impaired in cerebellar disease in humans. *Behav. Neurosci.* **108:** 1–11.

Cordo, P.J. 1988. Kinesthetic coordination of a movement sequence in humans. *Neurosci. Lett.* **92:** 40–45.

———. 1990. Kinesthetic control of a multijoint movement sequence. *J. Neurophysiol.* **63:** 161–172.

Cordo, P.J. and M. Flanders. 1989. Sensory control of target acquisition. *Trends Neurosci.* **12:** 110–117.

Darling, W.G. and G.F. Miller. 1993. Transformations between visual and kinesthetic coordinate systems in reaches to remembered object locations and orientations. *Exp. Brain Res.* **93:** 534–547.

Dow, R.S. and G. Moruzzi. 1958. *The physiology and pathology of the cerebellum.* University of Minnesota Press, Minneapolis, MN.

Flanders, M. and P.J. Cordo. 1989. Kinesthetic and visual control of a bimanual task: Specification of direction and amplitude. *J. Neurosci.* **9:** 447–453.

Gao, J.H., L.M. Parsons, J.M. Bower, J. Xiong, J. Li, and P.T. Fox. 1996. Cerebellum implicated in sensory acquisition and discrimination rather than motor control. *Science* **272:** 545–547.

Gilman, S., H. Carr, and J. Hollenberg. 1976. Kinematic effects of deafferentation and cerebellar ablation. *Brain* **99:** 311–330.

Gilman, S., J.R. Bloedel, and R. Lichtenberg. 1981. *Disorders of the cerebellum.* Davis Co., Philadelphia, PA.

Grill, S.E., M. Hallett, C. Marcus, and L. McShane. 1994. Disturbances of kinaesthesia in patients with cerebellar disorders. *Brain* **117:** 1433–1447.

Granit, R. 1970. *The basis of motor control.* Academic Press, London, UK.

Haines, D.E. (ed.) 1997. *Fundamental neuroscience.* Churchill Livingstone, New York.

Kawato, M. and H. Gomi. 1992. The cerebellum and VOR/OKR learning models. *Trends Neurosci.* **15:** 445–453.

Lovelace, E.A. 1989. Vision and kinesthesis in accuracy of hand movement. *Percep. Motor Skills* **68:** 707–714.

MacKay, W.A. and J.T. Murphy. 1979. Cerebellar modulation of reflex gain. *Prog. Neurobiol.* **13:** 361–417.

Nawrot, M. and M. Rizzo. 1995. Motion perception deficits from midline cerebellar lesions in human. *Vision Res.* **35:** 723–731.

Paulin, M.G. 1993. The role of the cerebellum in motor control and perception. *Brain Behav. & Evol.* **41:** 39–50.

Pellionisz, A.J. 1985. Tensorial brain theory in cerebellar modeling. In *Cerebellar functions* (ed. J.R. Bloedel, J. Dichgans, and W. Precht), pp. 201–229. Springer-Verlag, Berlin, Germany.

Pellionisz, A. and R. Llinas. 1985. Tensor network theory of the metaorganization of functional geometries in the central nervous system. *Neuroscience* **16:** 245–273.

Roll, J.P. and J.C. Gilhodes. 1995. Proprioceptive sensory codes mediating movement trajectory perception: human hand vibration-induced drawing illusions. *Can. J. Physiol. & Pharmacol.* **73:** 295–304.

Shimansky, Y., M. Saling, D.A. Wunderlich, V. Bracha, G.E. Stelmach, and J.R. Bloedel. 1996. Perception of kinesthetic cues required for assessing the shape of two-dimensional irregular profiles is impaired in cerebellar patients. *Neurosci. Abst.* **22:** 500.

Soechting, J.F. and M. Flanders. 1992. Moving in three-dimensional space: Frames of reference, vectors, and coordinate systems. *Annu. Rev. Neurosci.* **15:** 167–191.

Stein, J.F. and M. Glickstein. 1992. Role of the cerebellum in visual guidance of movement. *Physiol. Rev.* **72:** 967–1017.

Tillery, S.I., M. Flanders, and J.F. Soechting. 1991. A coordinate system for the synthesis of visual and kinesthetic information. *J. Neurosci.* **11:** 770–778.

Timmann, D., Y. Shimansky, P.S. Larson, D.A. Wunderlich, G.E. Stelmach, and J.R. Bloedel. 1996. Visuomotor learning in cerebellar patients. *Behav. Brain. Res.* **81:** 99–113.

Received March 4, 1997; accepted in revised form May 7, 1997.

Lateral Cerebellar Hemispheres Actively Support Sensory Acquisition and Discrimination Rather Than Motor Control

Lawrence M. Parsons,[1,3] James M. Bower,[2] Jia-Hong Gao,[1] Jinhu Xiong,[1] Jinqi Li,[1] and Peter T. Fox[1]

[1]Research Imaging Center
The Medical School
The University of Texas Health Science Center at San Antonio
San Antonio, Texas 78284
[2]Computation and Neural Systems
California Institute of Technology
Pasadena, California 91125

Abstract

This study examined a new hypothesis proposing that the lateral cerebellum is not activated by motor control per se, as widely assumed, but is engaged during the acquisition and discrimination of tactile sensory information. This proposal derives from neurobiological studies of these regions of the rat cerebellum. Magnetic resonance imaging of the lateral cerebellar output nucleus (dentate) of humans during passive and active sensory tasks confirmed four a priori implications of this hypothesis. Dentate nuclei responded to cutaneous stimuli, even when there were no accompanying overt finger movements. Finger movements not associated with tactile sensory discrimination produced no dentate activation. Sensory discrimination with the fingers induced an increase in dentate activation, with or without finger movements. Finally, dentate activity was greatest when there was the most opportunity to modulate the acquisition of the sensory tactile data: when the discrimination involved the active repositioning of tactile sensory surface of the fingers. Furthermore, activity in cerebellar cortex was strongly correlated with observed dentate activity. This distinct four-way pattern of effects strongly challenges other cerebellar theories. However, contrary to appearances, neither our hypothesis nor findings conflict with behavioral effects of cerebellar damage, neurophysiological data on animals performing motor tasks, or cerebellar contribution to nonmotor, perceptual, and cognitive tasks.

Introduction

The cerebellum has been viewed by most neurologists and neurobiologists in this century as principally a motor organ (Holmes 1917; Thach et al. 1992; Glickstein 1993; Ito 1993; Paulin 1993; Courchesne et al. 1994; Daum and Ackermann 1995; Dean 1995; Filapek 1995; Houk and Wise 1995; Welsh et al. 1995; Thach 1997). The most conspicuous and therefore best-studied effects of cerebellar lesions are on movement coordination (Holmes 1939; Thach et al. 1992; Subramony 1994; Gilman et al. 1996). Classic theory has emphasized that the output of the deep cerebellar nuclei in mammals is to primarily influence motor structures (Asanuma et al. 1983; Aumann et al. 1994; Keifer and Houk 1994). Cerebellar activity during movement has been well confirmed by electrophysiological recordings in animals (Ojakangas and Ebner 1992, 1994; Keifer and Houk 1994; van Kan et al. 1994; Apps et al. 1995) and by functional neuroimaging in humans (Fox et al. 1985; Ellerman et al. 1994). As a consequence, nearly all cellular and

[3]Corresponding author.

network level theories of cerebellar function focus on movement control (Marr 1969; Llinás 1984; Bloedel and Kelly 1992; Kawato and Gomi 1992; Paulin 1993; Houk and Wise 1995; Martin and Albers 1995). Despite the wealth of data concerning the possible involvement of the cerebellum in motor control, the specific computation executed by the cerebellum is far from established. For example, the role of the cerebellum in motor learning is a hotly contested issue, as demonstrated by the discussion in the present journal.

Be that as it may, the speculations on cerebellar function have recently been greatly diversified by numerous neurological, neuroimaging, and anatomical findings. Studies of brain-injured humans suggest that the cerebellum may be instrumental in nonmotor behaviors such as judging the timing of events, solving perceptual and spatial reasoning problems, and generating words according to a semantic rule (Bracke-Tolkmitt et al. 1989; Ivry and Keele 1989; Fiez et al. 1992; Grafman et al. 1992; Courchesne et al. 1994; Horak and Diener 1994). Support for some involvement in such perceptual and cognitive behaviors has also been obtained from recent neuroimaging studies of healthy humans (Petersen et al. 1989; Kim et al. 1994; Bower 1995; Jueptner et al. 1995; Leiner et al. 1995; Parsons et al. 1995). Furthermore, strong focused cerebellar activation has been detected for the mental rotation of abstract objects (L.M. Parsons and P.T. Fox, unpubl.), for different kinds of musical performance (P.T. Fox, L.M. Parsons, D.A. Hodges, and J.S. Sergent, in prep.), and for a variety of visual information processing tasks (Allen et al. 1997; Shulman et al. 1997). In these mental rotation, music, and visual studies, either motor behavior was not present in the task or activity owing to the task's motor components was dissociated from that owing to its nonmotor components. In addition, these functional data are consistent with recent anatomical studies that have discovered connectivity between cerebellum, midbrain, and nonmotor regions of the cerebral cortex (for recent discussion, see Middleton and Strick 1994; Schmahmann 1996, 1997).

Such findings clearly challenge classical motor theories of cerebellar function and have created an environment in which researchers are increasingly questioning the extent to which the cerebellum is involved in motor control. Although in most cases new alternative proposals are described as an extension of the role of the cerebellum in motor coordination itself, driven primarily by new neuroimaging results, others are proposing radically different functions for this ancient structure (for reviews, see Doyon 1997; Fiez and Raichle 1997; Parsons and Fox 1997).

This paper tests one such new theory of cerebellar function that posits that the cerebellum is specifically involved in monitoring and adjusting the acquisition of most of the sensory data on which the rest of the nervous system depends. Although space does not permit a complete presentation of the arguments for this hypothesis (Bower 1992, 1997a,b), we note briefly that it was suggested by results from electrophysiological studies of the tactilely responsive regions of the rat cerebellum (Bower and Woolston 1983; Paulin et al. 1989b; Sasaki et al. 1989; Bower and Kassel 1990; Gonzalez et al. 1993; Jaeger and Bower 1994). These regions of the cerebellum have extensive tactile representations in rats but no reported representation from proprioceptors (Welker 1987). Furthermore, these tactile representations are dominated by representations of the lip and whisker regions of the rat's face rather than by representations of the fore- or hindlimb, as would be expected from an involvement in limb movement coordination. Only selected regions of the rat's face are represented in these maps: those regions on the "leading edge of the rat" that are known to be in direct contact with objects during sensory exploration (Jacquin and Zeigler 1983). Furthermore, there is an unusual fractured somatotopic pattern of the tactile projections that seems to assure that data from the different regions of the rat's face involved in sensory exploration are juxtaposed in different combinations, perhaps reflecting behavioral relationships between the different body surfaces involved in tactile exploration (Jacquin and Zeigler 1983). Finally, a reanalysis of the physiological organization of cerebellar cortical circuitry has suggested that this network may be organized very differently than heretofore assumed (Jaeger and Bower 1994). This new understanding of cerebellar cortical organization is consistent with the idea that these regions of cerebellum control the use of these different surfaces during tactile sensory data acquisition (Bower 1997a).

Classical cerebellar theory, of course, has assumed that the cerebellum does not have sensory functions. There are other grounds for reconsidering this assumption however. The cerebellum receives input from virtually every sensory system (Brodal 1978; Brodal 1981; Welker 1987), including vestibular (Naito et al. 1995), proprioceptive

(Donga and Dessem 1993), visual (Snider and Stowell 1944), audition (Snider and Stowell 1944; Yang and Weisz 1992), somatosensation (Snider and Stowell 1944; Welker 1987), visceral pain receptors (Rubia and Phelps 1970), and sense pressure in blood vesicles (Nisimaru and Katayama 1995; Ghelarducci and Sebastiani 1996). In addition, it is activated by tactile stimulation alone (without overt movement) (Fox et al. 1985). Finally, determining whether a brain area has a motor or sensory function is a subtle, and often overlooked, problem. Motor behavior is guided by ongoing sensory acquisition of information about the object toward which action is directed. Continuously updated sensory data is necessary for accurate, coordinated, and smooth motor behavior. Nonetheless, we are convinced that it is necessary for motor operations to be disentangled as much as possible from sensory ones when testing hypotheses about cerebellar function.

Our study was designed specifically to dissociate tactile sensory data acquisition and discrimination from motor performance per se, by imaging blood flow change, a correlate of neural activity, in the lateral (dentate) nucleus of humans performing tasks both involving passive and active tactile discriminations. The dentate nucleus in humans is the sole output for the large lateral hemispheres of the primate cerebellum that are analogous to the region of tactile representation in the rat cerebellum (Larsell 1972) discussed above. In previous experimental studies of these regions in nonhuman primates, which are also known to respond to tactile stimuli, dentate and lateral hemisphere activation has been interpreted in the context of the control of finger movements for their own sake (Thach 1970, 1978; Strick 1983; van Kan et al. 1993). In an experimental study of this region in humans, Kim et al. (1994) used fMRI (functional magnetic resonance imaging) to assess activation in dentate nuclei during two tasks that were assumed to have comparable motor activity but different extents of "cognitive" processing (i.e., problem solving). The three times greater dentate activity detected during the more cognitive task was assumed to be attributable to nonmotor cognitive processing. However, this interpretation should be viewed with caution (Parsons and Fox 1997) because (1) only the more cognitive task was likely to involve imagined motor behavior and (2) imagined motor behavior is known to activate cerebellum as well as virtually all of the motor areas in cerebral cortex (Parsons et al. 1995).

In our study, we tested four implications of the alternative hypothesis that dentate activation will be more closely associated with sensory discriminations using the fingers than with finger-movement control in general. (1) Dentate nuclei should respond to sensory stimuli even when there are no accompanying overt finger movements. (2) Finger movements not associated with tactile sensory discrimination should not induce substantial dentate activation. (3) The requirement to make a sensory discrimination with the fingers should induce an increase in dentate activation, with or without accompanying finger movements. (4) The dentate should be most strongly activated when there is the most opportunity to modulate the acquisition of the sensory data, that is, when the sensory discrimination involves the active repositioning of tactile sensory surfaces through finger movements.

Materials and Methods

Six healthy volunteers performed four tasks. Subjects for the Cutaneous Stimulation (CS) and Cutaneous Discrimination (CD) tasks were males, ages 32–44 years (mean, 38). Five of those subjects and a female did the Grasped Objects (GO) and Grasped Objects Discrimination (GOD) tasks. All subjects gave informed consent. Subjects were given intensive practice session for 10 mins. In one session, they performed three cycles of Rest Control, CS, CD and, in another, three cycles of Rest Control, GO, GOD. Participants lay supine in a 1.9-T MRI instrument during each task. Subjects were blindfolded during all tasks and never saw the experimental stimuli.

In the CS Task, subjects passively experienced sandpaper rubbed against the immobilized pads of the second, third, and fourth fingers of each hand. In the CD Task, they were asked to actively compare (without responding) whether the coarseness of the sandpaper on their two hands matched. The coarseness of the sandpaper changed randomly every 3 sec. In the CS and CD tasks, movements were prevented by immobilization and instruction. The arm was immobilized by straps encircling the body. The wrist, hand, and fingers were immobilized by a rigid wooden surface affixed to the dorsum of the hand by tape encircling the wrist and fingers. During stimulation, subjects were instructed to allow their hand to remain flaccidly immobile, resting in the restraints. Sandpaper was applied to the finger pads by a continuous oscillation uncoordinated between hands. Four grades of sandpaper were used: 60, 100, 150, or 400 (U.S.A. Standard Grading system paper maximally packing grains of sand of 268, 141, 33, and 23.6 µm). As the sandpaper was mildly aversive, involuntary movements toward the stimuli were unlikely. Involuntary movements away from the stimuli were prevented by the rigid surface to which the fingers were attached. Subjects were visually monitored for movement throughout each trial. No movements were observed.

In the two grasping tasks, each (unrestrained) hand was enclosed in its own tightly woven cotton sock, each sock containing an identical set of four differently shaped stimuli. Each stimulus was a 2.5-mm-diameter wooden, smoothly surfaced sphere; the stimuli were differentiated by one-, two-, or three-faceted surfaces. In the GO Task, subjects were trained to pick up, roll in their fingers, and then release individual spheres with both sets of fingers. These instructions were intended to replicate movements made in the sensory discrimination as closely as possible, without a sensory discrimination actually being made. Subjects selected one of the four objects at random to grasp in each grasp cycle. This sequence was repeated continuously and was performed independently by each hand. In the GOD Task, subjects pincer-grasped a stimulus with one hand, felt its shape while using the other hand to grasp and feel another stimulus, and noticed whether the two objects were identical in shape. If the objects were different, the object in the right hand was dropped and another object was grasped and compared to the object in the left hand. If the objects matched, participants released both stimuli and immediately started the cycle over, beginning with the other hand than in the previous cycle.

During the prescan training sessions on the grasping tasks, subjects observed a sample performance of the tasks and then were blindfolded and performed each task without having their hands covered from view from the experimenters who closely monitored and coached their performance. During task performance in the fMRI sessions, subjects were also observed by the experimenters to ensure that the movements were consistent with those during training. The goal of the training and observation during scanning was to ensure that the movements in the two tasks were as comparable as possible. During the prescan training sessions on the cutaneous tasks, subjects observed a sample performance of the tasks and performed each task as experimenters closely monitored the performance to ensure that there was no finger, hand, limb, or body movement.

The fMRI images were made with a whole-body MRI system operating at 81 mHz (Gyrex, Elscint, Ltd., Israel). The subject's head was immobilized with a facial mask. A body coil was used for the radio-frequency transmission. The signal was received by a quadrature surface coil (US Asia Instruments, Inc., Highland Heights, OH). The dentate nuclei were functionally mapped with T_2^* gradient-echo images (Drayer et al. 1986). To locate the dentate nucleus for functional scanning, each session was started by using several high-resolution T2-weighted spin-echo images in the transverse plane perpendicular to the brain stem to cover the entire cerebellum. Two or three of the resulting images always contain the dentate nucleus, which is readily identified in these images because it is relatively dark (as a result of its high-iron-content dentate, it has low signal intensity). On the basis of the location of the dentate in the latter images, we selected the functional scanning image plane (of 6-mm thickness) to pass through the middle of the dentate. Human dentate nucleus is ~7–11 mm in the direction parallel to the brain stem, but most of its mass is concentrated in the middle (thinning out at the upper and lower ends). Thus, image planes we selected for recording functional activation are likely to contain much more than 60% of the dentate. Typically, the T_2^*-weighted images were collected by a conventional gradient-echo sequence with three cycles of the Rest Control and each of two tasks. Twenty images were acquired for each task within each cycle (7.8 sec per image). Imaging parameters were as follows: echo time, 40 msec; repetition time, 60 msec; and flip angle, 20°. Images were acquired with 256 complex pairs in the readout direction and 128 phase-encoding steps in a field of view 25.6×25.6 cm^2, with an in-plane spatial resolution of 1 by 2 mm^2. The acquired 128×256 fMRI data were zero-filled to 256×256, then Fourier-transformed. Maps of functional activation were calculated by comparison of T_2^*-weighted images acquired during the rest control with those acquired during tasks (Xiong et al. 1995).

Activation (Figs. 2 and 3, below) was defined by the combination of two criteria to afford a 0.05 level of statistical significance. First, a t-test was used to compare the resting baseline to task activation, and only pixels with a significant activation ($t > 2.5$) were included in the functional map. Second, regions with less than five contiguous activated pixels were excluded. This cluster analysis method addresses the multiple comparison problem while maintaining reasonable sensitivity (for discussion, see Friston et al. 1994; Forman et al. 1995; Xiong et al. 1995). This strategy of using two thresholds (intensity and area) is based on the fact that neural activity tends to cluster in spatial location and that MRI noise tends to be spatially uncorrelated. The resulting functional map was laid over a T_2^*-weighted image to determine activation sites. The area of activation used for comparisons was the number of activated pixels within the dentate nuclei. Omnibus statistics and planned comparisons were made with separate ANOVAs and Newman–Keuls tests of subjects' activation area values and intensity values for left and right dentate nuclei.

Results

All four predictions of the hypothesis were confirmed. Figure 1 shows the statistically significant, group-averaged, task-induced functional MRI (black) for each task laid onto group-averaged anatomical MRI (gray). Figures 2 and 3 quantify in bar graphs the extent and intensity, respectively, of the statistically significant, group-averaged, task-induced activation detected for each task.

The dentate nuclei showed significant task-induced increases in blood flow in the CS condition. Thus, dentate nuclei are activated by purely sensory stimuli, confirming positron-emission tomography results showing cerebellar activation during hand vibration (Fox et al. 1985). This activation was equally strong in the left and right dentate but tended to be more extensive in the right dentate, probably reflecting the left cerebral dominance of our right-handed subjects. Known anatomical connections (Brodal 1978; Glickstein et al. 1985; Schmahmann and Pandya 1995) may enable cerebellar participation in such sensory processing.

When the same stimuli were presented under identical conditions, but a discrimination was required (CD), dentate nuclei were more than twice

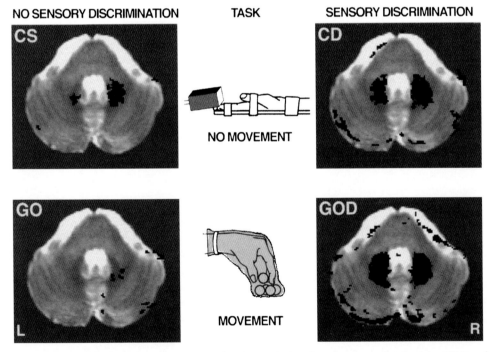

Figure 1: Functional MRI (black) laid onto anatomical MRI (gray) showing dentate activations for CS (*top left*), CD (*top right*), GO (*bottom left*), and GOD (*bottom right*) tasks. The dentate nuclei are the two dark crescent-shaped structures on either side of the cerebellar midline. Functional and anatomical images were coregistered for each task by performing rotation, translation, and scaling on each subject's images and then averaging images across subjects. A group *t*-test was performed on these images for each task comparing task-induced changes relative to rest. Activation was detected with a threshold defined by both a *t* value of 2.5 and a cluster of five adjacent pixels. The detected activations are statistically significant to $P < 0.05$ relative to the whole cerebellar plane sampled and shown in black scale ranging from a *t* value of 2.5 to 10.0.

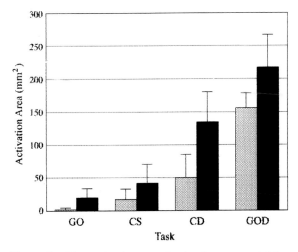

Figure 2: The activation area in the dentate nuclei for each task. (Shaded bar) Left dentate; (solid bar) right dentate. A group *t*-test was applied to each subject's data for each task compared with rest. Then, for each subject, activation foci were detected by selecting areas that had a *t* value >2.5 and at least five adjacent pixels (corresponding in combination to $P < 0.05$). The mean (±S.E.M.) of these activated foci was calculated for each task across subjects.

as active ($P < 0.05$). This activation was bilateral, but stronger in the right dentate. The enhanced activity could reflect the anatomical connections between these cerebellar regions and the prefrontal cortex that supports working memory processes (Baddeley 1986; Middleton and Strick 1994; Goldman-Rakic 1995) that are possibly necessary for discrimination. The activations observed in these two cutaneous tasks are thus consistent with the hypothesis that the cerebellum is engaged during the acquisition and processing of sensory information and is even more strongly engaged during discrimination. It is not yet known what mechanism or pathway is responsible for this increase in activity with the requirement to make a discrimination; we hypothesize that this increase reflects an anticipatory influence of the cerebral cortex on the cerebellum (Morissette and Bower 1996).

We also compared cerebellar activation in a sensory discrimination task requiring rapid coordinated finger movements to a control task with similar movements, but no discrimination. The control task (GO) produced very slight, statistically insignificant activation. The slight activation likely reflects cutaneous stimulation of the fingers touching the stimuli. The lack of activation in the GO Task confirms that rapid coordinated, fine finger movements, in the absence of a sensory discrimination,

do not engage the dentate nucleus. This response is equal to the slight dentate activity recorded in another fine motor behavior (Kim et al. 1994). In the latter study, the subject used visual guidance to reach to and grasp a peg, then raise and place it in the next small hole in linear array of such holes. The lack of dentate activation in such fine motor control tasks is opposite of what would be expected if existing theories of cerebellar motor control were correct and is evidence that the primary role of the cerebellum is not coordination of motor behavior for its own sake. The fact that active fine-grained finger movements do not alone significantly activate the dentate nuclei indicates that even if during either cutaneous task subjects made finger movements too slight to be detected by the experimenters, those movements per se would not cause the significant activations. This conclusion is further assurance that the dentate is indeed specifically activated by the acquisition and discrimination of tactile sensory information.

By far the strongest activation occurred when subjects made a covert discrimination of object

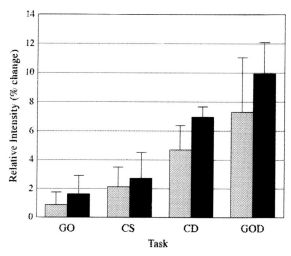

Figure 3: The intrinsic relative signal changes at the activation area in the dentate nuclei for each task. (Shaded bar) Left dentate; (solid bar) right dentate. A group *t*-test was applied to each subject's data for each task compared with rest. Then, for each subject, activation foci were detected by selecting areas that had a *t* value >2.5 and at least five adjacent pixels (corresponding in combination to $P < 0.05$). For each subject, the relative signal change was calculated in the above-threshold activation areas by subtracting the average signal value during rest from that during task and dividing by the average signal value during rest. Mean relative signal change (±S.E.M.) was calculated for each task across subjects.

shape using their fingers (GOD). Again, right dentate was slightly more active than in the left. There was >23 times as much activation observed when a discrimination was required in the grasping task than when it was not, and this extreme contrast ($P < 0.005$) documents strong cerebellar support for sensory discrimination. These data are consistent with earlier position emission tomography (PET) experiments reporting posterior cerebellar activity during tactile exploration tasks (Seitz et al. 1991).

Taken together, these results rule out the conclusion that the greater dentate activity in the GOD Task simply reflects fine motor control in general. The GO Task, which requires similarly fine motor control, produces no significant dentate activation. Thus, fine movement control per se does not engage the dentate, in contrast to sensory stimulation per se. The massive increase in activity in the GOD Task relative to that in the GO Task is entirely out of proportion to whatever subtle differences exist between the two tasks' very similar finger movements. The primary difference in movements—that GOD was performed at a slightly slower pace—would incorrectly predict a decrease in activation because motor performance rate and activation strength are positively correlated (Seitz et al. 1990; Sabatini et al. 1993). Thus, we are not claiming that the finger movements in GOD and GO were identical, just that both types of movements fall within the generally accepted category of "fine finger control" and that from the point of view of motor theories of cerebellum the enormous difference in activation would be completely unexpected.

There was also significant activation detected in the cortex of the lateral cerebellar hemispheres contained in the image plane (Fig. 1). Across the four conditions, activity in the lateral cerebellar cortex was positively correlated ($r = +0.91$, $P < 0.02$) with that in the dentate. The extent of statistically significant, task-induced, group-averaged activation detected in cerebellar cortex and dentate are quantified in the bar graph in Figure 4. The relationship between the dentate nucleus and cerebellar cortex is complex; nonetheless, this finding, although based on a small sample of cerebellar cortex, supports our hypothesis about the function of lateral hemispheres.

Discussion

In summary, our results consistently implicate the dentate nucleus of the human cerebellum in

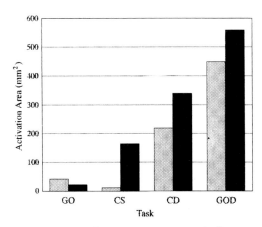

Figure 4: Bilateral activation in cerebellar cortex (shaded bar) and in dentate nuclei (solid bar) in the image plane sampled here. The extent of activation (in mm^2) is plotted for each of the four tasks used in this study.

sensory acquisition and discrimination. Activation occurred during sensory stimulation when there were no accompanying overt finger movements or discrimination. Substantial finger movements, when not associated with tactile discrimination, did not induce significant activation. Dentate activation was greatly enhanced when a sensory discrimination was required, with or without overt finger movements. However, the strongest activation occurred when sensory discrimination was combined with finger movements.

Although these findings implicate the lateral cerebellum in sensory discrimination rather than movement per se, they do not identify its specific role. For example, it is possible to interpret the greater increase in dentate activity for the grasping discrimination task as simply being attributable to the increased complexity of this sensory processing task compared with the pure cutaneous discrimination (Gibson 1962; Clark and Horch 1986; Lederman and Klatsky 1987; Parsons and Shimojo 1987; Reed et al. 1996). Discrimination based on the active manipulation of objects presumably involves a greater variety and amount of information to be acquired and integrated over time and space, including information about multiple finger positions, the three-dimensional spatial arrangement of object surfaces constituting shape, and tactile information about specific surfaces. In contrast, the cutaneous discrimination task involved passively forming a representation of a uniform surface composed of few dimensions of information (i.e., coarseness). The additional processing required to represent the more complex stimuli for the purpose of discrimination could be distributed among the somatosensory cortex, working memory areas in prefrontal cortex, and the cerebellum. As discussed earlier, there is appropriate connectivity for such distributed function. On the other hand, the interpretation of these differences in activation that is most in keeping with the hypothesis we set out to test is that the higher levels of cerebellar activation in the case of active manipulation reflect a direct role of the cerebellum in modulating the finger motor control system to reposition the tactile sensory surfaces of the fingers. This coordination may be based on the cerebellar analysis of the sense data actually being acquired and serves to assure that the highest quality information about object shape is being obtained in a coordinated fashion from all surfaces involved in the discrimination (Bower 1997a,b).

These findings are sharply contrary to the prevailing framework of cerebellar theory. Although space does not permit a full discussion of how one might comprehend or reconcile the contradictions, we will briefly describe below some of the more important issues involved. First, the data reported here are reliable. As noted in Results, the data in each separate condition are very similar to those of studies with other subjects, from other neuroimaging modalities, and by other research groups working with different experimental hypotheses and theoretical orientations. In addition, we have replicated the results using different subjects and a different imaging modality, PET, to measure activity throughout the whole cerebellum and cerebral cortex (L.M. Parsons, J.M. Bower, J. Xiong, M. Saenz, and P.T. Fox, unpubl.). Second, as described in the introduction, these tasks were designed to specifically test a new cerebellar theory previously developed over several years (Bower and Kassel 1990; Bower 1997a,b). The theory was not generated after the fact to fit the data. Third, as noted in Results, the data are internally consistent, each finding directly implied by our explicit hypothesis of cerebellar function. Fourth, motor control functions cannot explain the dentate and cerebellar cortex activation in response either to nonmotor sensory tasks or to the requirement to a make a sensory discrimination (see Results). Fifth, as described below, our hypotheses and findings are not inconsistent with principal observations of cerebellar research such as the behavioral effects of lesions; the possible role of cerebellum in associative learning; neurophysiological studies of non-human cerebellum in awake, behaving subjects; cerebellar participation in nonmotor, perceptual, and cognitive tasks; and the role of the cerebellar flocculus in regulation of the vestibulo-ocular reflex.

The hypotheses and findings reported here are not inconsistent with principal effects of cerebellar damage on behavior. For example, cerebellar deficits in voluntary movement such as incoordination and ataxia may result from disruption of the sensory data on which the motor system depends, rather than from disruption of cerebellar computations of smooth motor performance per se. Specifically, such deficits could be attributable to disruption of the known influence by the anterior cerebellum on the gamma motor system that is responsible for providing essential data to the motor system during motor control. Likewise, the decreased rate at which cerebellar patients perform motor tasks can be explained by the present hypothesis because longer computing time would be expected to process poorly controlled sensory data (for studies on patients' movement times, see Hallett et al. 1991; Wild and Dichgans 1992; Jahanshahi et al. 1993; Haggard et al. 1994; Muller and Dichgans 1994; Wessel et al. 1994; B. Wild and J. Dichgans, unpubl.). The same explanation accounts for the known increase in performance latencies for sensory-related tasks (Ivry and Keele 1989; Keele and Ivry 1990; Wessel et al. 1994; Tachibana et al. 1995). Furthermore, whereas many discriminations and behaviors can be completed in spite of the slowdown owing to the hypothesized lack of good sensory data, in tasks involving timing, any slowdown in processing time would be expected to interfere with task completion. For this reason, this hypothesis predicts that cerebellar dysfunction should have a noticeable effect on tasks involving timing, for which there is ample evidence (Ivry and Keele 1989; Keele and Ivry 1990; Lundyekman et al. 1991). Finally, the hypothesis would accommodate the suggested cerebellar involvement in autism (Courchesne et al. 1994; also see Filapek 1995) because a dysfunctional cerebellum would yield inconsistent or variable control of sensory data acquisition and would result in the reported difficulties autistic children have in dealing with sensory stimulation.

Our findings are also not incompatible with the data concerning some cerebellar involvement in associative learning. For example, several researchers have reported that specific cerebellar lesions interfere with the ability of rabbits to associate conditioned stimuli with eye blinks (Thompson and Krupa 1994). Although these experiments are controversial (Welsh and Harvey 1992; Bloedel and Bracha 1995; Logan and Grafton 1995), the point here is that none of those studies have taken into account the possibility that the sensory data on which the classical conditioning is dependent has changed. If cerebellar lesions do disrupt the structure of the sensory data, then one would also expect a disruption in conditioned responses, although it then becomes impossible to draw conclusions about the location of the "memory trace" from these experiments.

In addition, our results are not inconsistent with the many rigorous neurophysiological studies of awake behaving animals producing data interpreted as indicating that the cerebellum is involved in the planning and control of motor behavior. This view of the cerebellum is supported by recent trends in the types of tasks required of primates during cerebellar recording sessions. Specifically, early experiments of this type required monkeys to make large-scale controlled movements of the arms while the hands were in contact with a manipulandum (Thach 1970; Gilbert and Thach 1977; Harvey et al. 1977; Bauswein et al. 1983; Houk and Gibson 1987; Dugas and Smith 1992). These studies generally did not consider the possibility that recorded cerebellar responses might be more related to the changing tactile contact of the hand with the manipulandum. Similar experiments and interpretations have also been applied to cats (Schwartz et al. 1987; Lou and Bloedel 1992; Apps et al. 1995). More recently, researchers, largely driven by the desire to produce more robust cerebellar responses (cf. Houk and Gibson 1987; van Kan et al. 1994), have been training monkeys to make fine movements of the fingers that are increasingly like sensory discrimination tasks (Houk and Gibson 1987; Dugas and Smith 1992; van Kan et al. 1994; Robinson 1995; Sinkjaer et al. 1995). Although these studies are still interpreted in the context of motor control and do clearly have a motor component, they also require considerably more sensory processing than the old tasks involving the smooth movement of bars. Unfortunately, these animal studies (as well as other neuroimaging studies) did not attempt to decouple the sensory and motor processing as we have done here. As a result, it is plausible that cerebellar activity in the latter studies was also owing to acquisition of proprioceptive or tactile information rather than to computation involving motor coordination per se.

Our data are also compatible with the cerebellum participation in many varieties of sensory-motor, sensory, higher perceptual, and higher cognitive tasks (for reviews, see Fiez 1996; Bower 1997a,b; Parsons and Fox 1997; Schmahmann 1997; for related discussion see Bower 1995; Leiner et al. 1995). If the function of the cerebellum is to support the control of the acquisition of sensory data in general, not just tactile information, then it follows that the cerebellum should be active during a range of behaviors, including motor, perceptual, and cognitive tasks, all of which entail the acquisition of sensory information.

Finally, these data and our hypothesis about the role of cerebellar hemispheres in the somatosensory system are perhaps most consistent with the well-established role of the cerebellum in eye movement control. By definition, all eye movement control is directly related to sensory data acquisition, as this is all the eyes do. The specific role played by the cerebellum in eye movement control is perhaps best understood in the floccular regions of the cerebellum involved in the regulation of the vestibulo-ocular reflex. This reflex specifically compensates for the dislocation of images on the retina as a result of body movement (Paulin et al. 1989a,b). It is already known that the flocculus receives information signaling the occurrence of retinal slip during head movements (Ito et al. 1982) and is presumed to be involved in reducing the amount of retinal slip. Although the cerebellum effects this control through the extraocular eye muscles (Ito et al. 1982), the computational objective of the cerebellar control is to reduce retinal slip, which is known to degrade the processing of visual information in the visual cortex (Westheimer and McKee 1975). The flocculus does not take part in the discrimination of visual objects, but it increases the efficiency of function of the visual system. The present hypothesis proposes a comparable involvement for the lateral cerebellar hemispheres in tactile data acquisition. Interestingly, the flocculus is one of the phylogenetically oldest cerebellar regions, whereas the cerebellar hemispheres are the youngest (Larsell 1972).

In conclusion, the contribution of the cerebellum to sensory acquisition and discrimination must be investigated further to be clearly understood. Experiments in preparation have studied these same four tasks while neuroimaging the activity in the red and dentate nuclei (Y. Pu, J.-H. Gao, Y.J. Liu, M. Liotti, P.T. Fox, J.M. Bower, and L.M. Parsons, unpubl.) and while neuroimaging the whole brain and whole cerebellum, using correlated activation to evaluate hypotheses about distributed function (L.M. Parsons, J.M. Bower, J. Xiong, M. Saenz, and P.T. Fox, unpubl.). More generally, as the variety of behavior in which the cerebellum is reported to participate increases, it becomes critical to disentangle the component elementary operations of each behavior and determine where and to what extent each operation is selectively supported within the cerebellum. The benefit of such a strategy is demonstrated in evidence now emerging for regional dissociation of function (for review, see, e.g., Parsons and Fox 1997). At the same time, a thorough understanding of cerebellar function will depend on developing a better understanding of the computational significance of cerebellar cortical and deep nuclear circuitry (see, e.g., De Schutter and Maex 1996; Jaeger et al. 1997). Finally, as our grasp of the principles of distributed computation deepens, hypotheses about cerebellar function are likely to have quite a different form and be supported by quite different arguments, than at present.

Acknowledgments

This work was supported by an EJLB Foundation grant and National Institute of Mental Health grants P20 DA52176-01 and MH/DA52145 (Human Brain Project), and MH/DA52145-04 (Human Brain Project) to J.M.B.

The publication costs of this article were defrayed in part by payment of page charges. This article must therefore be hereby marked "advertisement" in accordance with 18 USC section 1734 solely to indicate this fact.

References

Allen, G., R.B. Buxton, E.C. Wong, and E. Courchesne. 1997. Attentional activation of the cerebellum independent of motor involvement. *Science* **275:** 1940–1943.

Apps, R., N.A. Hartell, and D.M. Armstrong. 1995. Step phase-related excitability changes in spino-olivocerebellar paths to the c(1) and c(3) zones in cat cerebellum. *J. Physiol. (Lond.)* **483:** 687–702.

Asanuma, C., W.T. Thach, and E.G. Jones. 1983. Brain-stem and spinal projections of the deep cerebellar nuclei in the monkey, with observations on the brain-stem projections of the dorsal column nuclei. *Brain Res. Rev.* **5:** 299–322.

Aumann, T.D., J.A. Rawson, D.I. Finkelstein, and M.K. Horne. 1994. Projections from the lateral and interposed cerebellar nuclei to the thalamus of the rat—a light and electron-microscopic study using single and double anterograde labeling. *J. Comp. Neurol.* **349:** 165–181.

Baddeley, A. 1986. *Working memory.* Oxford University Press, New York, NY.

Bauswein, E., F.P. Kolb, B. Leimbeck, and F.J. Rubia. 1983. Simple and complex spike activity in cerebellar Purkinje cells during active and passive movements in the awake monkey. *J. Physiol. (Lond.)* **339:** 379–394.

Bloedel, J.R. and T.M. Kelly. 1992. The dynamic selection hypothesis: A proposed function for cerebellar sagittal zones. In *The cerebellum revisited* (ed. R.R. Llinás and C. Sotelo), pp. 267–282. Springer-Verlag, New York, NY.

Bloedel, J.R. and V. Bracha. 1995. On the cerebellum, cutaneomuscular reflexes, movement control, and the elusive engrams of memory. *Behav. Brain Res.* **68:** 1–44.

Bower, J.M. 1992. Is the cerebellum a motor control device? *Behav. Brain Sci.* **15:** 714–715.

———. 1995. The cerebellum as a sensory acquisition controller: Commentary on "The underestimated cerebellum" by Leiner et al. *Hum. Brain Mapping* **2:** 255–256.

———. 1997a. Is the cerebellum sensory for motor's sake, or motor for sensory's sake: The view from the whiskers of a rat? In *Progress Brain Research* (ed. C.I. de Zeeuw, P. Strata, and J. Voogd), pp. 483–516. Elsevier Science, Amsterdam, The Netherlands.

———. 1997b. Control of sensory data acquisition. In *The cerebellum and cognition* (ed. J.D. Schmahmann). Academic Press, New York, NY.

Bower, J.M. and D.C. Woolston. 1983. Congruence of spatial organization of tactile projections to granule cell and Purkinje cell layers of cerebellar hemispheres of the albino rat: Vertical organization of cerebellar cortex. *J. Neurophysiol.* **49:** 745–766.

Bower, J.M. and J. Kassell. 1990. Variability in tactile projection patterns to cerebellar folia crus-IIa of the Norway rat. *J. Comp. Neurol.* **302:** 768–778.

Bracke-Tolkmitt, R., A. Linden, G.M. Canavan, B. Rockstroh, E. Scholz, K. Wessell, and H.C. Diener. 1989. The cerebellum contributes to mental skills. *Behav. Neurosci.* **103:** 442–446.

Brodal, A. 1981. *Neurological anatomy in relation to clinical medicine.* Oxford University Press, New York, NY.

Brodal, P. 1978. The corticopontine projection in the rhesus monkey: Origin and principles of organization. *Brain* **101:** 251–283.

Clark, F.J. and K.W. Horch. 1986. Kinethesia. In *Handbook of perception and human performance: Vol. 1. sensory processes and perception* (ed. K.R. Boff, L. Kaufman, and J.P. Thomas). Wiley, New York, NY.

Courchesne, E., J. Townsend, N.A. Akshoomoff, O. Saitoh, R. Yeung-Courchesne, A.J. Lincoln, H.E. James, R.H. Haas, L. Schreibman, and L. Lau. 1994. Impairment in shifting attention in autistic and cerebellar patients. *Behav. Neurosci.* **108:** 848–865.

Daum, I. and H. Ackermann. 1995. Cerebellar contributions to cognition. *Behav. Brain Res.* **67:** 201–210.

Dean, P. 1995. Modeling the role of the cerebellar fastigial nuclei in producing accurate saccades—The importance of burst timing. *Neuroscience* **68:** 1059–1077.

De Schutter, E. and R. Maex. 1996. The cerebellum: Cortical processing and theory. *Curr. Opin. Neurobiol.* **6:** 759–764.

Donga, R. and D. Dessem. 1993. An unrelayed projection of jaw-muscle spindle afferents to the cerebellum. *Brain Res.* **626:** 347–350.

Doyon, J. 1997. Skill learning. In *The cerebellum and cognition* (ed. J.D. Schmahmann). Academic Press, New York, NY.

Drayer, B., P. Burger, R. Darwin, S. Riederer, R. Herfkens, and G.A. Johnson. 1986. MRI of brain iron. *Am. J. Roentgenol.* **147:** 103–110.

Dugas, C. and A.M. Smith. 1992. Responses of cerebellar Purkinje cells to slip of a hand-held object. *J. Neurophysiol.* **67:** 483–495.

Ellerman, J.M., D. Flament, S.-G. Kim, Q.-G. Fu, H. Merkle, T.J. Ebner, and K. Ugurbil. 1994. Spatial patterns of functional activation of the cerebellum investigated using high-field (4-t) MRI. *Magnetic Resonance Imaging Biomed.* **7:** 63–68.

Fiez, J.A. 1996. Cerebellar contributions to cognition. *Neuron* **116:** 13–15.

Fiez, J. and M.E. Raichle. 1997. Linguistic processing. In *The cerebellum and cognition* (ed. J.D. Schmahmann). Academic Press, New York, NY.

Fiez, J.A., S.E. Petersen, M.K. Cheney, and M.E. Raichle. 1992. Impaired nonmotor learning and error-detection associated with cerebellar damage—a single case study. *Brain* **115:** 155–178.

Filapek, P.A. 1995. Quantitative magnetic-resonance imaging in autism—the cerebellar vermis. *Curr. Opin. Neurol.* **8:** 134–138.

Fox, P.T., M.E. Raichle, and W.T. Thach. 1985. Functional mapping of the human cerebellum with positron emission tomography. *Proc. Natl. Acad. Sci.* **82:** 7462–7466.

Friston, K.J., P. Jezzard, and R. Turner. 1994. Analysis of functional MRI time-series. *Hum. Brain Mapping* **1:** 153–171.

Ghelarducci, B. and L. Sebastiani. 1996. Contribution of the cerebellar vermis to cardiovascular control. *J. Auton. Nerv. Syst.* **56:** 149–156.

Gibson, J.J. 1962. Observations on active touch. *Psychol. Rev.* **69:** 477–491.

Gilbert, P.F.C. and W.T. Thach. 1977. Purkinje cell activity during motor learning. *Brain Res.* **128:** 309–328.

Gilman, S., A.A.F. Sima, L. Junck, K.J. Kluin, R.A. Koeppe, M.E. Lohman, and R. Little. 1996. Spinocerebellar ataxia type-1 with multiple system degeneration and glial cytoplasmic inclusions. *Ann. Neurol.* **39:** 241–255.

Glickstein, M. 1993. Motor skills but not cognitive tasks. *Trends Neurosci.* **16:** 450–451.

Glickstein, M., J. May, and B. Mercier. 1985. Corticopontine projection in the macaque: The distribution of labeled cortical cells after large injections of horseradish peroxidase in the pontine nuclei. *J. Comp. Neurol.* **235:** 343–359.

Goldman-Rakic, P.S. 1995. Toward a circuit model of working memory and the guidance of voluntary motor action. In *Models of information processing in the basal ganglia* (ed. J.C. Houk, J.L. Davis, and D.G. Beiser), pp. 131–148. Massachusetts Institute of Technology Press, Cambridge, MA.

Gonzalez, L., C. Shumway, J. Morissette, and J.M. Bower. 1993. Developmental plasticity in cerebellar tactile maps—fractured maps retain a fractured organization. *J. Comp. Neurol.* **332:** 487–498.

Grafman, J., I. Litvan, S. Massaquoi, M. Stewart, A. Sirigu, and M. Hallett. 1992. Cognitive planning deficits in patients with cerebellar atrophy. *Neurology* **42:** 1493–1496.

Haggard, P., J. Jenner, and A. Wing. 1994. Coordination of aimed movements in a case of unilateral cerebellar damage. *Neuropsychologia* **32:** 827–846.

Hallett, M., A. Berardelli, J. Matheson, J. Rothwell, and C.D. Marsden. 1991. Physiological analysis of simple rapid movements in patients with cerebellar deficits. *J. Neurol. Neurosurg. Psychiatry* **54:** 124–133.

Harvey, R.J., R. Porter, and J.A. Rawson. 1977. The natural discharges of Purkinje cells in paravermal regions of lobules V and VI of the monkey's cerebellum. *J. Physiol. (Lond.)* **271:** 515–536.

Holmes, G. 1917. The symptoms of acute cerebellar injuries due to gunshot wounds. *Brain* **40:** 461–535.

———. 1939. The cerebellum of man. *Brain* **62:** 1–30.

Horak, F.B. and H.C. Diener. 1994. Cerebellar control of postural scaling and central set in stance. *J. Neurophysiol.* **72:** 479–493.

Houk, J.C. and A.R. Gibson. 1987. Sensorimotor processing through the cerebellum. In *New concepts in cerebellar neurobiology* (ed. J.S. King), pp. 387–416. A.R. Liss, New York, NY.

Houk, J.C. and S.P. Wise. 1995. Distributed modular architectures linking basal ganglia, cerebellum, and cerebral-cortex—their role in planning and controlling action. *Cereb. Cortex* **5:** 95–110.

Ito, M. 1993. Movement and thought—identical control mechanisms by the cerebellum. *Trends Neurosci.* **16:** 448–450.

Ito, M., M. Sakurai, and P. Tongroach. 1982. Climbing fibre induced depression of both mossy fibre responsiveness and glutamate sensitivity of cerebellar Purkinje cells. *J. Physiol. (Lond.)* **324:** 113–134.

Ivry, R.B. and S.W. Keele. 1989. Timing functions of the cerebellum. *J. Cognit. Neurosci.* **1:** 136–152.

Jacquin, M.F. and H.P. Zeigler. 1983. Trigeminal orosensation and ingestive behavior in the rat. *Behav. Neurosci.* **97:** 62–97.

Jaeger, D. and J.M. Bower. 1994. Prolonged responses in rat cerebellar Purkinje-cells following activation of the granule cell layer—An intracellular in-vitro and in-vivo investigation. *Exp. Brain Res* **100:** 200–214.

Jaeger, D., E. De Schutter, and J.M. Bower. 1997. The role of synaptic and voltage-gated currents in the control of Purkinje cell spiking: A modeling study. *J. Neurosci.* **17:** 91–106.

Jahanshahi, M., R.G. Brown, and C.D. Marsden. 1993. A comparative-study of simple and choice-reaction time in parkinsons, huntingtons and cerebellar disease. *J. Neurol. Neurosurg. Psychiatry* **56:** 1169–1177.

Jueptner, M., M. Rijntjes, C. Weiller, H.H. Faiss, D. Timmann, S.P. Mueller, and H.C. Diener. 1995. Localization of a cerebellar timing process using PET. *Neurology* **45:** 1540–1545.

Kawato, M. and H. Gomi. 1992. A computational model of 4 regions of the cerebellum based on feedback-error learning. *Biol. Cybern.* **68:** 95–103.

Keele, S.W. and R. Ivry. 1990. Does the cerebellum provide a common computation for diverse tasks—a timing hypothesis. *Ann. N.Y. Acad. Sci.* **608:** 179–211.

Keifer, J. and J.C. Houk. 1994. Motor function of the cerebellorubrospinal system. *Physiol. Rev.* **74:** 509–542.

Kim, S.-G., K. Ugurbil, and P.L. Strick. 1994. Activation of a cerebellar output nucleus during cognitive processing. *Science* **265:** 949–951.

Larsell, O. 1972. *The comparitive anatomy and histology of the cerebellum from monotremes through apes.* University of Minnesotta Press, Minneapolis, MN.

Lederman, S.J. and R.L. Klatsky. 1987. Hand movements: A window into haptic object recognition. *Cognit. Psychol.* **19:** 342–368.

Leiner, H.C., A.L. Leiner, and R.S. Dow. 1995. The underestimated cerebellum. *Hum. Brain Map.* **2:** 244–254.

Llinás, R. 1984. Functional significance of the basic cerebellar circuit in motor coordination. In *Cerebellar functions* (ed. J.R. Bloedel), pp. 171–185. Springer-Verlag, New York, NY.

Logan, C.G. and S.T. Grafton. 1995. Functional anatomy of human eyeblink conditioning determined with regional cerebral glucose metabolism and positron-emission tomography. *Proc. Natl. Acad. Sci.* **92:** 7500–7504.

Lou, J.-S. and J.R. Bloedel. 1992. Responses of sagittally aligned Purkinje cells during perturbed locomotion: Relation of climbing fiber activation to simple spike modulation. *J. Neurophysiol.* **68:** 1820–1833.

Lundyekman, L., R. Ivry, S. Keele, and M. Woollacott. 1991. Timing and force control deficits in clumsy children. *J. Cognitive Neurosci.* **3:** 367–376.

Marr, D. 1969. A theory of cerebellar cortex. *J. Physiol. (Lond.)* **202:** 437–471.

Martin, P. and M. Albers. 1995. Cerebellum and schizophrenia—a selective review. *Schizophr. Bull.* **21:** 241–250.

Middleton, F.A. and P.L. Strick. 1994. Anatomical evidence for cerebellar and basal ganglia involvement in higher cognitive function. *Science* **266:** 458–461.

Morissette, J. and J.M. Bower. 1996. The contribution of somatosensory cortex to responses in the rat cerebellar granule cell layer following peripheral tactile stimulation. *Exp. Brain Res.* **109:** 240–250.

Muller, F. and J. Dichgans. 1994. Dyscoordination of pinch and lift forces during grasp in patients with cerebellar lesions. *Exp. Brain Res.* **101:** 485–492.

Naito, Y., A. Newman, W.S. Lee, K. Beykirch, and V. Honrubia. 1995. Projections of the individual vestibular end-organs in the brain-stem of the squirrel-monkey. *Hear. Res.* **87:** 141–155.

Nisimaru, N. and S. Katayama. 1995. Projection of cardiovascular afferents to the lateral nodulus-uvula of the cerebellum in rabbits. *Neurosci. Res.* **21:** 343–350.

Ojakangas, C.L. and T.J. Ebner. 1992. Purkinje-cell complex and simple spike changes during a voluntary arm movement learning-task in the monkey. *J. Neurophysiol.* **68:** 2222–2236.

———. 1994. Purkinje-cell complex spike activity during voluntary motor learning—relationship to kinematics. *J. Neurophysiol.* **72:** 2617–2630.

Parsons, L.M. and S. Shimojo. 1987. Perceived spatial organization of cutaneous patterns on surfaces of the human body in various positions. *J. Exp. Psychol.* **13:** 488–504.

Parsons, L.M. and P.T. Fox. 1997. Sensory and cognitive functions. In *The cerebellum and cognition* (ed. J.D. Schmahmann), pp. 255–271. Academic Press, New York, NY.

Parsons, L.M., P.T. Fox, J.H. Downs, T. Glass, T.B. Hirsch, C.C. Martin, P.A. Jerabek, and J.L. Lancaster. 1995. Use of implicit motor imagery for visual shape discrimination as revealed by PET. *Nature* **375**: 54–59.

Paulin, M.G. 1993. The role of the cerebellum in motor control and perception. *Brain Behav. Evol.* **41**: 39–50.

Paulin, M.G., M.E. Nelson, and J.M. Bower. 1989a. Dynamics of compensatory eye movement control: An optimal estimation analysis of the vestibulo-ocular reflex. *Int. J. Neural Syst.* **1**: 23–29.

———. 1989b. Neural control of sensory acquisition: The vestibulo-ocular reflex. In *Advances in neural information processing systems* (ed. D. Touretzky), Vol. 1, pp. 410–418. Morgan Kaufmann, San Mateo, CA.

Petersen, S.E., P.T. Fox, M.I. Posner, M. Mintun, and M.E. Raichle. 1989. Positron emission tomographic studies of the processing of single words. *J. Cognit. Neurosci.* **1**: 153–170.

Reed, C.L., R.J. Caselli, and M.J. Farah. 1996. Tactile agnosia: Underlying impairment and implications for normal tactile object recognition. *Brain* **119**: 875–888.

Robinson, F.R. 1995. Role of the cerebellum in movement control and adaptation. *Curr. Opin. Neurobiol.* **5**: 755–762.

Rubia, F.J. and J.B. Phelps. 1970. Responses of the cerebellar cortex to cutaneous and visceral afferents. *Pflügers Arch.* **314**: 68–85.

Sabatini, U., F. Chollet, O. Rascol, P. Celsis, A. Rascol, G.L. Lenzi, and J.-P. Marc-Vergnes. 1993. Effect of side and rate of stimulation on cerebral blood flow changes in motor areas during finger movements in humans. *J. Cereb. Blood Flow Metab.* **13**: 639–645.

Sasaki, K., J.M. Bower, and R. Llinás. 1989. Multiple purkinje-cell recording in rodent cerebellar cortex. *Eur. J. Neurosci.* **1**: 572–586.

Schmahmann, J.D. 1996. From movement to thought: Structural and functional correlates of the cerebellar contribution to cognitive processing. *Hum. Brain Mapping* **4**: 174–198.

Schmahmann, J.D., ed. 1997. *The cerebellum and cognition.* Academic Press, New York, NY.

Schmahmann, J.D. and D.N. Pandya. 1995. Prefrontal cortex projections to the basilar pons: Implications for the cerebellar contribution to higher function. *Neurosci. Lett.* **199**: 175–178.

Schwartz, A.B., T.J. Ebner, and J.R. Bloedel. 1987. Responses of interposed and dentate neurons to perturbations of the locomotor cycle. *Exp. Brain Res.* **67**: 323–338.

Seitz, R.J., P.E. Roland, C. Bohm, T. Greitz, and S. Stone-Elander. 1990. Motor learning in man. *NeuroReport* **1**: 57–60.

———. 1991. Somatosensory discrimination of shape—tactile exploration and cerebral activation. *Eur. J. Neurosci.* **3**: 481–492.

Shulman, G.L., M. Corbetta, R.L. Buckner, J.A. Fiez, F.M. Miezin, M.E. Raichle, and S.E. Petersen. 1997. Common blood flow changes across visual tasks: I. Increases in subcortical structures and cerebellum, but not in non-visual cortex. *J. Cognit. Neurosci.* (in press).

Sinkjaer, T., L. Miller, T. Andersen, and J.C. Houk. 1995. Synaptic linkages between red nucleus cells and limb muscles during a multijoint motor task. *Exp. Brain Res.* **102**: 546–550.

Snider, R.S. and A. Stowell. 1944. Receiving areas of the tactile, auditory, and visual systems in the cerebellum. *J. Neurophysiol.* **7**: 331–357.

Strick, P.L. 1983. The influence of motor preparation on the response of cerebellar neurons to limb displacements. *J. Neurosci.* **3**: 2007–2020.

Subramony, S.H. 1994. Degenerative ataxias. *Curr. Opin. Neurol.* **7**: 316–322.

Tachibana, H., K. Aragane, and M. Sugita. 1995. Event-related potentials in patients with cerebellar degeneration—electrophysiological evidence for cognitive impairment. *Cognit. Brain Res.* **2**: 173–180.

Thach, W.T. 1970. Discharge of cerebellar neurons related to two maintained postures and two prompt movements: Purkinje cell output and input. *J. Neurophysiol.* **33**: 537–547.

———. 1978. Correlation of neural discharge with pattern and force of muscular activity, joint position, and direction of intended next movement in motor cortex and cerebellum. *J. Neurophysiol.* **41**: 654–676.

———. 1997. Cerebellum, motor learning and thinking in man: Pet studies. *Behav. Brain Sci.* (in press).

Thach, W.T., H.P. Goodkin, and J.G. Keating. 1992. The cerebellum and the adaptive coordination of movement. *Annu. Rev. Neurosci.* **15**: 403–442.

Thompson, R.F. and D.J. Krupa. 1994. The organization of memory traces in the mammalian brain. *Annu. Rev. Neurosci.* **17**: 519–549.

van Kan, P.L.E., J.C. Houk, and A.R. Gibson. 1993. Output organization of intermediate cerebellum in monkey. *J. Neurophysiol.* **69**: 57–73.

van Kan, P.L.E., K.M. Horn, and A.R. Gibson. 1994. The importance of hand use to discharge of interpositus neurons of the monkey. *J. Physiol. (Lond.)* **480**: 171–190.

Welker, W. 1987. Spatial organization of somatosensory projections to granule cell cerebellar cortex: Functional and connectional implications of fractured somatotopy. In *New* concepts in cerebellar neurobiology (ed. J.S. King), pp. 239–280. Liss, New York, NY.

Welsh, J.P. and J.A. Harvey. 1992. The role of the cerebellum in voluntary and reflexive movements. In *The cerebellum revisited* (ed. R.R. Llinás and C. Cotelo), pp. 301–334. Springer-Verlag, New York, NY.

Welsh, J.P., E.J. Lang, I. Sugihara, and R. Llinás. 1995. Dynamic organization of motor control within the olivocerebellar system. *Nature* **374:** 453–457.

Wessel, K., R. Verleger, D. Nazarenus, P. Vieregge, and D. Kompf. 1994. Movement-related cortical potentials preceding sequential and goal-directed finger and arm movements in patients with cerebellar atrophy. *Electroencephalogr. Clin. Neurophysiol.* **92:** 331–341.

Westheimer, G. and S.P. McKee. 1975. Visual acuity in the presence of retinal-image motion. *J. Opt. Soc. Am.* **65:** 847–850.

Xiong, J., J.-H. Gao, J.L. Lancaster, and P.T. Fox. 1995. Clustered pixels analysis for functional MRI activation studies of the human brain. *Hum. Brain Mapping* **3:** 287–301.

Yang, B.Y. and D.J. Weisz. 1992. An auditory conditioned-stimulus modulates unconditioned stimulus elicited neuronal-activity in the cerebellar anterior interpositus and dentate nuclei during nictitating-membrane response conditioning in rabbits. *Behav. Neurosci.* **106:** 889–899.

Received February 14, 1997; accepted in revised form May 2, 1997.

Cerebellar Guidance of Premotor Network Development and Sensorimotor Learning

Sherwin E. Hua and James C. Houk[1]

Department of Physiology
Northwestern University Medical School
Chicago, Illinois 60611-3008

Abstract

Single unit and imaging studies have shown that the cerebellum is especially active during the acquisition phase of certain motor and cognitive tasks. These data are consistent with the hypothesis that particular sensorimotor procedures are acquired and stored in the cerebellar cortex and that this knowledge can then be exported to the cerebral cortex and premotor networks for more efficient execution. In this article we present a model to illustrate how the cerebellar cortex might guide the development of cortical–cerebellar network connections and how a similar mechanism operating in the adult could mediate the exportation of sensorimotor knowledge from the cerebellum to the motor cortex. The model consists of a three-layered recurrent network representing the cerebello-thalamocortical-ponto-cerebellar limb premotor network. The cerebellar cortex is not explicitly modeled. Our simulations show that Hebbian learning combined with weight normalization allows the emergence of reciprocal and modular structure in the limb premotor network. Reciprocal connections allow activity to reverberate around specific loops. Modularity organizes the connections into specific channels. Furthermore, we show that cerebellar learning can be exported to motor cortex through these modular and reciprocal premotor circuits. In particular, we simulate developmental alignment of visuomotor relations and their realignment as a consequence of prism exposure. The exportation of sensorimotor knowledge from the cerebellum to the motor cortex may allow faster and more efficient execution of learned motor responses.

Introduction

It is now well established that the cerebellum plays an important role in sensorimotor learning (Thompson 1986; Ito 1989; Thach et al. 1992; Houk et al. 1996; Raymond et al. 1996). Although cerebellar patients are able to make a wide range of movements, their ability to adapt to challenging conditions is severely impaired (Thach et al. 1992). In addition to basic movement skills, imaging studies show that the cerebellum is involved in complex sensorimotor functions that include cognitive components (Kim et al. 1994; Raichle et al. 1994; Friston et al. 1992).

Although the parallel fiber–Purkinje cell (PC) synapse is undoubtedly one important site at which learning takes place (Marr 1969; Albus 1971; Ito 1989; Berthier et al. 1993), learning mechanisms are widespread, and sensorimotor learning is undoubtedly a distributed function (Houk and Barto 1992; Bloedel et al. 1996). The storage capacity of the cerebellar cortex is quite enormous (Gilbert 1974; Tyrrell and Willshaw 1992), so one possibility is that a diversity of complex sensorimotor memories are stored in the cerebellum and that during practice, these programs are exported to premotor networks for more efficient execution (Galiana 1986; Houk and Barto 1992). Consistent with this hypothesis, single unit activity in the cerebellum is particularly intense during the acquisition phase of conditioned forelimb movements (Milak et al. 1995). Similarly, imaging studies indicate the most intense cerebellar activations when sensorimotor tasks are being learned (Friston et al. 1992; Ebner et al. 1996). Cognitive functions of the cerebellum could similarly be exported to the cerebral cortex to improve

[1]Corresponding author.

the efficiency of thinking (Houk 1997). Positron emission tomography (PET) studies showed that metabolic activity associated with a cognitive verb-finding task moved from the lateral cerebellum to a cortical site with practice (Raichle et al. 1994). These results, along with previous studies showing that cerebellar damage hinders normal motor learning but not the retention of motor memories, suggest that the cerebellar cortex can export learning of motor and cognitive tasks to premotor and cortical sites.

Although cerebellar learning has been modeled extensively, mechanisms whereby cerebellar knowledge might be exported to premotor and cortical sites have not received much attention. Here we explore this issue with respect to limb movement control. The premotor network that controls voluntary movements of the limb consists of a cerebello-thalamo-cortico-ponto-cerebellar circuit regulating motor cortical output and a cerebello-rubro-reticulo-cerebellar brain stem circuit regulating red nucleus output (for review, see Houk et al. 1993). Experiments have shown that the brain stem circuit exhibits reverberatory activity generated by recurrent positive feedback in the network (Tsukahara et al. 1983; Keifer et al. 1992; Keifer 1996). Workers using modeling efforts have assumed that the limb premotor network is comprised of reciprocal and topographically specific connections that segregate the control of movement into computational modules, each of which generates an elemental motor command (Houk et al. 1990; Berthier et al. 1993). Such specificity in the premotor circuits may allow activity patterns to be readily shaped by inhibitory output from cerebellar PCs to coordinate populations of motor cortical and red nucleus commands (Houk et al. 1993). Developmental plasticity leading to the formation of the postulated premotor modules might represent a neonatal phase of the adult plasticity involved in the exportation process.

In addition to the formation of the premotor network, developmental alignment of premotor circuits with cortico-cortical connections may guide the organization of sensorimotor maps in the cortex. For example, monkeys deprived of sight of their hands and body during development showed difficulty making and adapting visually guided movements later in life (Bauer and Held 1971). Thus, the establishment of these visuo-motor maps during development may be important for normal visuo-motor adaptation guided by cerebellar and premotor circuits later in life.

In this study we provide a computational model that first simulates the development of specific reciprocal circuits in the premotor network. Later in development, these reciprocal circuits guide the alignment of visuo-motor maps in cortico-cortical connections. After these two developmental stages are completed, the model simulates the exportation of cerebellar learning to cortico-cortical networks for a visuo-motor adaptation task analogous to prism adaptation of limb movements. To simplify the model, cerebellar learning is not explicitly modeled. Previous modeling efforts have dealt with cerebellar learning of visuomotor relations (Berthier et al. 1993). Instead, we focus on plasticity in premotor and cortical networks that is guided by the cerebellum.

The centerpiece of the model is the cerebello-thalamo-cortico-ponto-cerebellar recurrent network. Because the thalamo-cortical circuit is itself a recurrent network within the larger recurrent premotor network, we simplify the network by incorporating the recurrent thalamo-cortical circuit into a single layer of units having lateral connections and self-connections. The self-connections on this layer represent specific thalamo-cortical reciprocal loops that allow cortical activity to persist through thalamic connections. Lateral connections represent intrinsic cortical connectivity as well as limited divergence of thalamo-cortical recurrent connections. The incorporation of the thalamo-cortical circuit into a single layer is justified by experimental evidence demonstrating topographic specificity in reciprocal thalamo-cortical connections (Ma and Juliano 1991; Ghosh et al. 1994). Additionally, we have previously demonstrated that any two-layered recurrent network will develop reciprocal and symmetric connections through Hebbian learning (S.E. Hua, F.A. Mussa-Ivaldi, and J.C. Houk, unpubl.). The reciprocal connections in such two-layered networks can be approximated by a single composite weight matrix constructed by the product of the feed-forward and feedback weight matrices in the network. For topographically organized reciprocal networks, the composite weight matrix has a banded diagonal structure, which is equivalent to having self-weights and lateral connections.

Our simulations show that Hebbian learning during the early developmental stage allows the formation of reciprocal and modular connections in the premotor network. Reciprocity exists when activity initiated in a unit will return to that unit after traveling through multisynaptic connections in the network (self-connections are reciprocal but are not considered in this study). Modularity exists when a group of cells in the network has stronger connections with cells in the same group than with cells in other groups. We also define computational or graded modularity as a subclass of modularity in which cells have a uniformly graded distribution of connections that varies with functional or topographic similarity. For example, cortical units that have stronger connections with units having similar directional tuning or orientation selectivity have graded modularity.

Development of the premotor network is guided by spontaneous PC disinhibition of neighboring nuclear cells. Once development of the premotor network has occurred, the cerebellum is able to direct the alignment of visuo-motor maps in the cortex. Furthermore, the cerebellum can direct learning and realignment of cortico-cortical circuits by exporting cerebellar learning to cortical sites. After learning, cortico-cortical connections in our model are able to initiate movements in the correct direction in a visuo-motor adaptation task without direct control by the cerebellum.

Materials and Methods

The model premotor network is comprised of a three-layered recurrent network representing the cerebellar nucleus, the motor cortical–thalamic circuit, and the pontine nucleus, as schematized in Figure 1 (CBN, MC, and PN, respectively). A fourth layer representing cortical regions that specify visual target information (sensory cortex) projects to the motor cortex. These cortico-cortical connections are assumed to develop after connections in the premotor network have organized. Each layer of units is wrapped, meaning the ends were connected, to reduce border effects. Thus, effectively, each layer is circular in topology.

In the cerebellar nucleus layer, we assume for simplicity that each PC inhibits a cluster of three neighboring nuclear cells. This topographic arrangement allows better visualization of the effects of correlated PC disinhibition through the topography of resulting premotor connections. During the development of the premotor loop, nuclear cells have spontaneous activity, such that when inhibition from a PC is removed, the cluster of nuclear cells that received its inhibition becomes active. Later, after the premotor network has developed, we assume that cerebellar nuclear cells

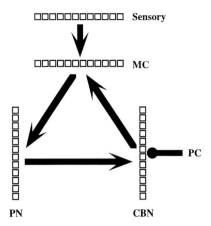

Figure 1: Schematic diagram of the network. The three-layered premotor network is comprised of the motor cortico-thalamo-cortical (MC) layer, the pontine (PN) layer, and the cerebellar nuclear (CBN) layer. Adaptable cortico-cortical connections exist between the sensory layer and the MC layer. PC inhibition regulates the CBN layer.

are no longer spontaneously active. Instead, excitatory pontine inputs are necessary in addition to PC disinhibition to activate nuclear cells. No inhibitory interneurons are included in the cerebellar nuclear layer.

Cerebellar nuclear activation is transmitted to the motor cortex via thalamic connections. For simplicity the reciprocal cortical–thalamic circuit is represented by a single layer of units (MC layer) with lateral excitatory connections and self connections. Self-connections represent reciprocal cortico-thalamic connections, whereas lateral connections represent both the intrinsic cortical connections as well as divergent cortico-thalamo-cortical connections. Additionally, inhibitory interneurons are included to stabilize cortical activity.

The pontine layer is a relay layer without external inputs other than the MC layer and without intrinsic lateral excitatory connections. Inhibitory interneurons are included on this layer. Histological studies have demonstrated the existence of principal relay cells that project to the cerebellum as well as inhibitory interneurons in the pontine nucleus (Brodal and Bjaalie 1992).

We assume that cortico-cortical connections develop after the basic organization of the premotor loop has been laid down. After this point, we fix the connections in the premotor loop and allow cerebellar learning to direct the alignment and realignment of the feed-forward connections from the sensory layer to the motor cortex. The sensory layer represents cortical regions such as parietal cortex or ventral premotor cortex that provide information about visual targets to which movements are to be made. The spatially correlated external inputs representing visual target information are implemented in the model as a cluster of three neighboring units on the sensory layer. Additionally, a corresponding PC is turned off resulting in the disinhibition of three neighboring cerebellar nuclear cells. The concurrent activation of premotor cortical units and disinhibition of cerebellar nuclear cells drives the formation of an appropriate cortico-cortical mapping.

Two dynamics occur simultaneously in the network, namely activation dynamics and synaptic efficacy or learning dynamics (S.E. Hua, F.A. Mussa-Ivaldi, and J.C. Houk, in prep.). The activation states of all units are updated asynchronously and are governed by the additive model,

$$x_j(t+dt) = x_j(t) + \left[-\frac{1}{R}x_j(t) + \sum_i W_{ji}(t)\sigma[x_i(t)] + b_j(t) \right]\frac{dt}{C}$$

where $x_j(t)$ represents the instaneous activation state $\sigma[x_j(t)]$ is the activation function whose value is comparable to firing rate, $W_{ji}(t)$ is the synaptic weight from unit i to unit j, RC is the time constant for activation dynamics, and $b_j(t)$ represents the time-varying inputs.

The activation rule used in the simulations is piecewise and is similar to Anderson's brain-state-in-a-box model (Anderson et al. 1977):

$$\sigma(x) = \begin{cases} 1 & \text{if } x \geq 1 \\ x & \text{if } 0 < x < 1 \\ 0 & \text{if } x \leq 0 \end{cases}$$

Learning occurs by $dW_{ji}(t) = \frac{\alpha}{\tau_w}\sigma[x_j(t)]\sigma[x_i(t)]dt$, the general Hebbian update rule where α is the learning rate and τ_w is the time constant for continuous learning. Becuase Hebbian learning in this form is generally unstable (Miller and MacKay 1994), we impose both presynaptic and postsynaptic normalization as a means of stabilizing Hebbian learning.

$$W_{ji}(t+dt) = \beta[W_{ji}(t)]_{post}[W_{ji}(t) + dW_{ji}(t)]\beta[W_{ji}(t)]_{pre}$$

where the coefficient for presynaptic normalization is defined by

$$\beta[W_{ji}(t)]_{pre} = \frac{X_{pre}}{\sum_b [W_{bi}(t) + dW_{bi}(t)]}$$

and the coefficient for postsynaptic normalization is defined by

$$\beta[W_{ji}(t)]_{post} = \frac{\chi_{post}}{\sum_b [W_{jb}(t) + dW_{jb}(t)]}$$

X_{pre} is the limit on the sum of all weights from each presynaptic unit, and X_{post} is the limit on the sum of all weights onto each postsynaptic unit. For both normalization conditions to be satisfied, the condition $X_{pre} = X_{post}$ must hold. (For presynaptic normalization, we have $\sum_{j=1}^{N} W_{ji} = \chi_{pre}$. For postsynaptic normalization we have $\sum_{i=1}^{N} W_{ji} = \chi_{post}$. Because $\sum_{i=1}^{N}\sum_{j=1}^{N} W_{ji} = N\chi_{pre}$ and $\sum_{j=1}^{N}\sum_{i=1}^{N} W_{ji} = N\chi_{post}$, it follows that $\chi_{post} = \chi_{pre}$ must hold if both normalization conditions are to be satisfied.) Although both learning and activation dynamics evolve simultaneously, the dynamics for weight update are slower than that for changes in the instantaneous firing rate, that is, $RC \ll \tau_w$. Typically, $RC = 1$, $\tau_w = 1000$, $dt = 0.1$, and $\alpha = 1$.

In addition to the excitatory projection neurons described above, linear inhibitory interneurons are included in the cortical and pontine layers. Each interneuron inhibits only one projection neuron with a weight of −1 and receives excitatory input from all projection units on the previous layer with a weight equal in magnitude to the mean of all excitatory weights of projection units. We assume that inhibitory interneurons have fast dynamics such that the time delay through the interneurons is negligible compared with the direct excitatory connections. Finally, noise is added to the weights at each time step at ±10% of the value of the largest weight. This noise represents spontaneous synaptic sprouting and elimination.

Results

The influence of cerebellar output on the learning and development of downstream circuits is simulated in three stages. In the first stage, initially random connections in the recurrent premotor network are organized into specific reciprocal circuits. Cerebellar activation during this developmental stage allows the alignment of specific PCs with specific reciprocal channels in the premotor circuit. In the second stage, premotor circuits are assumed to have passed the critical period of development and are no longer plastic. Visuo-motor maps in cortico-cortical projections are aligned by simultaneous presentation of visual target input to both the sensory layer and the cerebellar cortex. We assume that the cerebellar cortex has already learned how to command movements to these visual targets. In the third stage, cerebellar knowledge about prism distortion is exported to cortico-cortical connections through activation of specific premotor pathways.

DEVELOPMENT OF THE PREMOTOR CIRCUIT

The feedback circuit consists of a cerebellar nuclear layer, a cortical-thalamic layer, and a pontine layer as schematized in Figure 1. Each of the three projections in this circuit was initialized with random connections. The only external input to the network at this stage was PC inhibition of nuclear units. During development, we assumed that PCs are spontaneously active with intermittent periods of inactivity. We simulated intermittent PC inactivity by activating neighboring clusters of cerebellar nuclear cells for 75 time steps, a conservative number of steps to ensure network settling.

Under these developmental conditions, the connections in the feedback circuit developed reciprocal connections, as shown in Figure 2. Figure 2, a–c, shows the weight matrices for the cerebello-MC, MC-pontine, and ponto-cerebellar projections, respectively. Figure 2, d–f, shows the corresponding composite matrices for the cerebellar, MC, and pontine layers, respectively. Each composite matrix shows the effective connections that a layer makes with itself through the feedback connections in the network and is calculated by the product of the three individual weight matrices, taken in reverse order. The diagonal terms of the composite matrix represent the effective connection that a single unit makes with itself through the network; thus the diagonal terms give a measure of the degree of reciprocity in the network.

Figure 2a shows that the cerebello-MC projection developed a smooth topographic map. The MC-pontine and ponto-cerebellar projections developed partial topographic maps. The MC-pontine projection, shown in Figure 2b, has topographic receptive fields but discontinuous projection fields, such that neighboring MC units (labeled as layer 2) project to similar, topographically discontinuous pontine units. On the other hand, the ponto-cerebellar projection, shown in Figure 2c, has topographic projection fields but discontinuous receptive fields, such that neighboring cerebellar units have similar but discontinuous receptive fields.

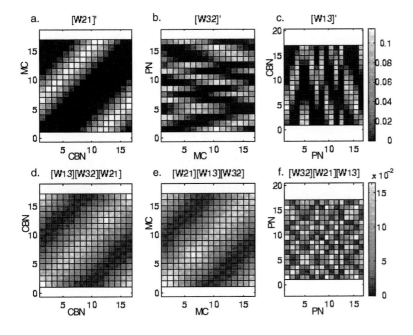

Figure 2: Weight matrices (a–c) and composite matrices (d–f) of the premotor network after the first developmental phase of Hebbian learning. (a) The weight matrix W_{21} from the cerebellar layer (layer 1) to the MC layer (layer 2); (b) the weight matrix W_{32} from the MC layer to the pontine layer (layer 3); (c) weight matrix W_{13} from the pontine layer to the cerebellar layer; (d) the composite matrix for the cerebellar layer, constructed by the product of the three weight matrices $W_{13}W_{32}W_{21}$; (e) the composite matrix for the MC layer, i.e., $W_{21}W_{13}W_{32}$; (f) the composite matrix for the pontine layer, i.e., $W_{32}W_{21}W_{13}$.

These weight maps are consistent with the correlation structure between individual units on each of the three layers. Both the MC and cerebellar layers have topographically correlated activity. Correlation of MC units arises from lateral connections between neighboring units. On the other hand, correlation of cerebellar nuclear cells is generated by correlated disinhibition by each PC. Units on the pontine layer do not have any correlation in activity imposed by the network architecture. Thus, the cerebello-MC projection between two layers with correlated activity develops a fully topographic map. The MC-pontine projection from a layer with correlated activity to a layer without correlations develops topographic receptive fields but nontopographic projection fields. The ponto-cerebellar projection from a layer without correlations to a layer with correlated activity develops topographic projection fields but non-topographic receptive fields.

As defined above, the composite matrix shows the multisynaptic connections that a unit makes with all other units on the same layer. The composite matrix for the cerebellar layer (Fig. 2d) is banded diagonal. This matrix structure shows that activity initiated in a cerebellar unit will travel through the other two layers and will return not only to that unit, by reciprocal connections as illustrated by the diagonal terms, but will also travel to neighboring cerebellar units, as exemplified by off-diagonal terms. The composite matrix for the MC, shown in Figure 2e, has a similar banded diagonal structure. In contrast, the composite matrix for the pontine nucleus, shown in Figure 2f, is a symmetric but otherwise disorganized matrix with prominent diagonal terms.

Once again, these composite matrices are consistent with the correlation structure between individual units on each layer. First, all three composite matrices have prominent diagonal terms. Diagonal terms on a composite matrix represent reciprocal connections that allow activity to travel through the network and return to the units from which the activity originated. Second, all three composite matrices are symmetric. Symmetry shows that the influence of unit A on unit B is the same as that of unit B on unit A through multisynaptic connections in the network. Finally, for layers with correlated activity, the composite matrix is banded, representing a propensity for recurrent activity to return not only to units that initiated the activity but also to neighboring units that have correlated activity. In contrast, the pontine nucleus does not have correlated activity, and the organization of its composite matrix is less obvious.

ALIGNMENT OF CORTICAL VISUO-MOTOR MAPS

In this section we describe a model of the developmental alignment of cortico-cortical connections guided by cerebellar cortical activity. Visual target information is simultaneously provided to the sensory layer and to PCs in the cerebellar cortex (Fig. 1). We assume that the connections from sensory to MC layer are initially random, whereas the cerebellar cortex has already learned to use this sensory input to guide movements to appropriate endpoints. Experimental studies of the thalamo-cortical projection supports a critical period during development, after which plasticity is reduced (Iriki et al. 1991; Crair and Malenka 1995; Fox et al. 1996), and on this basis we assume that the plasticity of intrinsic premotor connections is turned off after the maturation of these connections. Cortico-cortical connections, however, remain plastic throughout development and adulthood (Iriki et al. 1991). Stable premotor connections allow the cerebellum to export motor learning to cortico-cortical sites by way of reliable connections.

In our model we presented visual target inputs to the sensory cortex and cerebellar cortex concurrently. For example, if units $n-1$, n, and $n+1$ were activated at the target layer, cerebellar units $n-1$, n, and $n+1$ were simultaneously disinhibited. All other cerebellar units were continually inhibited by full PC inhibition. These concurrent inputs were presented for 75 time steps and were chosen randomly. Our implementation of concurrent target input is only a simplification of the actual pathway by which these two areas receive concurrent inputs. In reality, visual target input is relayed to the cerebellum by way of the corticopontine and the mossy fiber projection to the cerebellum (Brodal and Bjaalie 1992).

Additionally, after the first developmental stage during which the basic architecture of the premotor weights are established, the gain of the network is allowed to increase. Electrophysiologic evidence shows that immature cortical neurons are less responsive than mature units (Prince and Huguenard 1988). Additionally, an increase in gain may occur by the maturation of neuromodulatory inputs such as cholinergic or noradrenergic inputs (McCormick 1989). We model a growth in network gain by increasing the size of the normalization constant χ from 0.5 to 5.0 after the first developmental stage. This growth in weights effectively increases the gain of the input–output transformation performed by each unit, allowing activity to reverberate in the circuit.

Figure 3 shows that the feed-forward projection between the sensory cortex and the MC layer developed a continuous topographic map. Figure 3a shows the initial random cortico-cortical map. After paired target inputs were given to the target layer and cerebellar layer, the cortico-cortical projection, shown in Figure 3b, developed a topographic map that corresponds closely to the cerebello–MC projection.

VISUO-MOTOR LEARNING

After the development and alignment of both the premotor circuit and the cortico-cortical projection, the network was subjected to a visual perturbation task in which the input to the visual cortex is shifted by four units. This shift in visual input is analogous to a prism adaptation task in which subjects wear prism goggles that shift visual space. A shift in visual inputs causes a shift in the pattern of PC deactivation, resulting in a movement error. We assumed that this movement error allows the cerebellum to learn the new visuo-motor mapping so that the correct movement can be made, compensating for the shift in visual target input. A previous modeling study has demonstrated that long-term depression at the parallel fiber synapse on PCs in an adjustable pattern generator (APG) model can simulate the reprogramming of movements to shifted sensory inputs (Berthier et al. 1993). In the present model, we assumed that the cerebellum has learned the new sensorimotor programming. Thus, even though visual target input is shifted at the sensory cortex, cerebellar cortical activity has learned to compensate by disinhibiting the unshifted pattern of cerebellar nuclear cells, allowing the correct movement to be initiated. Thus, shifted visual target input is paired with an unshifted disinhibition of cerebellar nuclear cells because of cerebellar learning of the new mapping. For example, if units $n - 1$, n, and $n + 1$ were activated at the sensory layer, cerebellar units $n + 3$, $n + 4$, and $n + 5$ received PC disinhibition, whereas all other cerebellar units were continually inhibited by full PC inhibition. The paired inputs were presented for 75 time steps and were chosen randomly.

Figure 3c shows that the feed-forward projection between the sensory layer and the MC layer developed a continuous and shifted topographic map. When compared with the topographic map in the nonshifted case (Fig. 3b), the topographic map in the shifted case is appropriately shifted by four units. Thus, activation of the appropriate output by the cerebellum allows the appropriate adaptation by cortico-cortical connections.

ACTIVATION DYNAMICS DURING DEVELOPMENT AND LEARNING

So far we have discussed only the pattern of weights that arise during development and learning. However, it is important to look at the behavior of the activation dynamics that occur with each pattern of connections. The activation dynamics at three stages of development are shown. First we show the behavior of the network after the first stage when the premotor network has fully developed but when the cortico-cortical connections

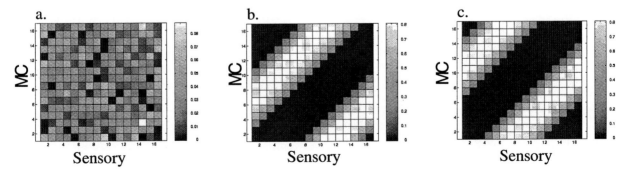

Figure 3: Development and adaptation in the cortico-cortical projection. (*a*) The initially random weight matrix; (*b*) the topographically organized cortico-cortical projection formed after the second developmental stage of visuo-motor alignment; (*c*) the shifted cortico-cortical projection after adaptation of a visuo-motor task in the learning phase.

were still random, as shown in Figure 3a. Second, we show that after the second stage when the cortico-cortical connections have formed, motor activity can be initiated and driven by cortico-cortical connections alone. Finally, we show that after cerebellar learning of the visuo-motor task has been exported to the cortex, cortico-cortical connections were able to initiate motor activity in the appropriate direction to compensate for the shifted visual input.

Figure 4 shows the dynamic behavior of a network after the first developmental stage. The premotor network has formed reciprocal connections, and the weights have grown in size so that the gain of the network has increased. The projection from the sensory cortex to the MC has not yet developed and still has random connections. Figure 4,

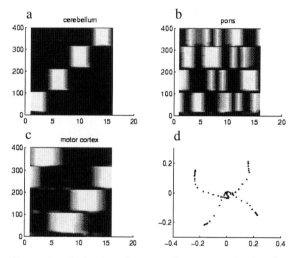

Figure 4: Activation dynamics for a network after the first developmental stage. This network has a mature premotor circuit but random cortico-cortical connections (Fig. 3a). (a–c) Activation dynamics over time of the cerebellar, pons, and MC layers when four inputs are given to the sensory cortex. Inputs are given to neighboring clusters of three units centered on units 2, 7, 11, and 14. Each input is held on for 105 time steps concomitant with disinhibition of the corresponding cerebellar units. The unit number is represented on the horizontal axis, whereas the number of time steps is represented on the vertical axis. (d) Temporal evolution of the population direction vector for each of the four inputs given to the network. Each dot represents the population vector calculated at a particular time step. The population vector is calculated by a vector summation of the individual direction vectors for each unit. For each unit the amplitude of its individual vector equals its firing rate, whereas the preferred direction is calculated by the following equation (unit number/16)*2*PI.

a–c, shows the activation states of the cerebellar, pontine, and MC layers, respectively, when four sets of input were sequentially presented to both the sensory cortex and the cerebellum. In these dynamic simulations, each of the four inputs was turned on for 105 time steps to fully demonstrate that the network settles to a steady state. The cerebellar layer shows four blocks of activation that correspond to the four sets of input, each causing disinhibition of a cluster of cerebellar nuclear cells. After the presentation of each new input, activity in the MC layer was initially disorganized. However, after activity had traveled through the premotor loop, the appropriate pattern of PC inhibition allowed premotor activity, and thus motor cortical activity, to be reshaped correctly.

Figure 4d shows the population direction vector in cartesian coordinates for each of the four inputs. Each point represents an instantaneous population vector. Because the ends of each layer are connected, the 16 units on each layer can each represent movement in a particular direction on a two-dimensional plane spanning 360°. The population vector is calculated by the spatially weighted average of the activations of the 16 units on the MC layer, with the direction specifity of each unit being distributed about circular layer in 22.5° intervals. Figure 4d shows that the population vectors from each of the four inputs were initially disordered but became directed at the four appropriate movement directions after PC inhibition had reshaped the activity in the premotor network.

Figure 5 shows the behavior of a network after the second developmental stage when the cortico-cortical projection has aligned with premotor circuits. In this simulation we removed cerebellar disinhibition, such that the visual target input is only given to the sensory cortex. This arrangement simulates a network where motor learning has already been exported to cortical sites, and cerebellar control is no longer needed. Thus, the movement can be initiated directly by cortico-cortical connections. The cerebellar nuclear layer, in Figure 5a, is silent throughout the sequential presentation of four sets of inputs. The MC layer shows four distinct blocks of activity corresponding to the correct sequence of motor activity as specified by the sensory cortical layer. The pontine layer shows clustered discontinuous activity as a result of MC activation. Figure 5d shows the population direction vectors of the MC layer. The four distinct movement directions correspond to the four different target inputs given to the sensory cortex. Note

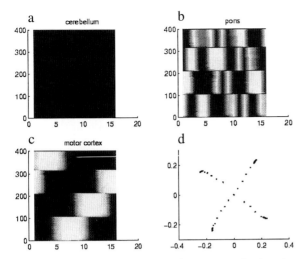

Figure 5: Activation dynamics for a network after the second developmental stage. The network has a mature premotor circuit and an aligned cortico-cortical map (Fig. 3b). All conventions are the same as that in Fig. 4, except no cerebellar disinhibition occurs in this simulation.

that with each input, the population vector initially grows in amplitude, and then the population vector remains fixed at a particular direction and amplitude.

Figures 6 and 7 show the behavior of a network that is exposed to a visuo-motor adaptation task. Figure 6 shows the activation dynamics after the cerebellum has learned to compensate for the 4-unit shift in visual target space but before exportation of the learning to cortical sites. Because the cortico-cortical connections have not adapted to the shift in target space, visual inputs to the sensory layer initially activate inappropriate motor units in the MC by direct cortico-cortical connections. However, because the cerebellum has learned to compensate for visual shift, the MC is able to attain the correct pattern of activation by PC shaping of premotor activity, shown in Figure 6c. Figure 6d shows the shift or rotation of the direction vector that occurs when cerebellar inhibition guides the behavior of MC activity.

After cerebellar learning of the visuo-motor task has been exported to the cortex, the cortico-cortical projection directs the appropriate pattern of MC activity in response to the shifted target inputs. Figure 7 shows a simulation where exportation of learning has already taken place, and the cerebellum no longer responds to the visual input. Figure 7a shows that the cerebellar nucleus has no activity. Instead, the sensory cortex is able to di-

rectly activate the correct pattern of MC units to compensate for the shift in visual target space. Figure 7d shows that the population direction vector grows directly in the correct direction through the realigned cortico-cortical connections.

Discussion

These modeling studies demonstrate that cerebellar output could be used to guide the development of premotor circuits as well as to direct sensorimotor learning in cortico-cortical connections. Spontaneous PC disinhibition with rebound activation of cerebellar nuclear units allowed the formation of reciprocal connections in the premotor network. These reciprocal connections permit activity to persist and reverberate in specific computational modules. After development of the premotor network is completed, paired target information to a visual sensory area of cortex and to the cerebellar cortex allowed the alignment of cortico-cortical maps. When visual target information is misaligned with movement, mimicking prism adaptation, cerebellar learning of the misalignment directs realignment of cortico-cortical connections. After realign-

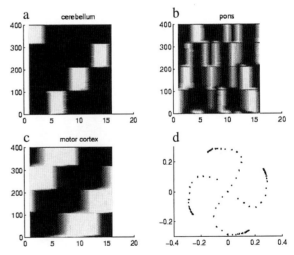

Figure 6: Activation dynamics for a network after the second developmental stage but subjected to a visuo-motor task. The network has a mature premotor circuit and an aligned cortico-cortical map (Fig. 3b). Inputs to the sensory cortex are shifted four units from the correct target. We assumed that the cerebellum has learned this visuo-motor mapping and disinhibits nuclear cells that are shifted by four units from the inputs given to sensory cortex. All other conventions are the same as that in Fig. 4.

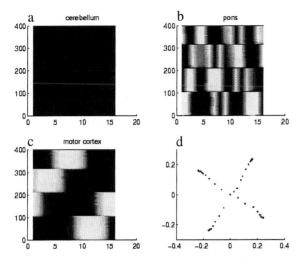

Figure 7: Activation dynamics for a network that has learned the visuo-motor task. The premotor circuit remains unchanged, but the cortico-cortical map is shifted as shown in Fig. 3c. Inputs to the sensory cortex are shifted by four units from the correct target. A corresponding shift in cortico-cortical connections alone, without disinhibition in the cerebellum, can compensate for the shift in visual inputs. All other conventions are the same as that in Fig. 4.

ment of cortical connections is accomplished, cortical areas supplying visual target information are able to initiate movement in the appropriate direction to move toward a target without further involvement of the cerebellum. Elimination of cerebellar involvement is advantageous, as it shortens the control pathway for responses that are used frequently, thus forming what has been referred to as "sensorimotor habits" (Houk and Barto 1992).

MODELING ASSUMPTIONS

Synaptic plasticity simulated in the present model resulted from Hebbian learning coupled with weight normalization. Hebbian learning is thought to be implemented by the NMDA channel (Brown et al. 1990) and has been demonstrated in several regions of the CNS including the cerebral cortex (Bear and Kirkwood 1993). Labeling studies have demonstrated NMDA labeling in cortex, thalamus, pons, and cerebellar nuclei (Petralia et al. 1994). These studies suggest that Hebbian learning could occur at each of the layers in the present model.

Labeling studies have shown that NMDA receptor subunits have differential expression patterns during development (Zhong et al. 1995). Thus, different levels of plasticity could occur during development. In the present model we assumed that plasticity in the premotor network occurred during early development. Several experimental studies show that the thalamo-cortical connection has a critical period for synaptic plasticity after which plasticity is reduced (Fox et al. 1996). Such a critical period has been demonstrated for several sensory cortical areas (Fox and Zahs 1994). In addition, Iriki et al. (1991) have shown that adult thalamo-cortical connections to motor cortex are less plastic when compared with cortico-cortical connections to the motor cortex. Although these studies strongly suggest that thalamo-cortical synapses become less plastic after early development when compared with cortico-cortical synapses, little is known about synaptic plasticity at other synapses in the premotor circuit. Further studies are needed to confirm the assumption that premotor synapses become less plastic after early development.

Hebbian learning is generally unstable because it only specifies a weight increment. Most models of recurrent networks include a weight decay mechanism to limit learning (Shinomoto 1987). In the present model, we used dual weight normalization to stabilize synaptic learning. Forms of weight normalization exist biologically at the level of nuclei (Hayes and Meyer 1988a,b) and at the level of neurons (Markram and Tsodyks 1996). Furthermore, most models of Hebbian learning in feed-forward networks utilize weight normalization (Whitelaw and Cowan 1981; Miller et al. 1989; Miller 1994). Such models can simulate the development of feed-forward pathways such as topographic maps, ocular dominance columns, and orientation domains. We have shown recently that weight normalization with Hebbian learning is effective in recurrent networks (S.E. Hua, F.A. Mussa-Ivaldi, and J.C. Houk, unpubl.). In that study weight normalization in a two-layered recurrent network allowed the formation of symmetric, reciprocal, and modular connections.

During the transition from the first to the second developmental phases, we assumed that synaptic weights in the premotor network grew in strength. This growth allows the mature premotor circuit to have sufficient network gain to sustain reverberatory activity. This up-regulation in the strength of synapses was accomplished by an increase in the normalization constant. There are two mechanisms that could accomplish such a transition in weight. First, during development, weights could undergo a gradual increase in the size of connections or neuronal responsiveness as neurons mature. Immature cortical neurons have been found to be less responsive than adult neurons (Prince and Huguenard 1988). Second, an increase in responsiveness can be achieved by modulatory neural transmitters. Cholinergic and adrenergic input have been shown to have both inhibitory and excitatory effects on cortical excitability (McCormick 1989). Both of these mechanisms could exist to allow an increased network gain after premotor connections are formed.

During the late developmental and learning phases, visual target information is provided simultaneously to the sensory cortical layer and to PCs. During the prism adaptation task, rotated target information is again provided to both layers, but the PCs are assumed to have learned the correct counter-rotation to achieve the appropriate movement. For simplicity, we did not explicitly include cerebellar learning in this model. Instead, we assumed that the cerebellum would be able to learn this counter-rotation to achieve the proper movement in spite of rotated target information. Previously, we have shown that an APG model of cerebellar function can learn the alignment between target and movement direction as well as adapt to visual distortion (Berthier et al. 1993). This model is consistent with the Marr–Albus theory of parallel fiber–PC synapse modification by error signals from climbing fibers (Marr 1969; Albus 1971). In the present model, learning at the cerebellar cortex is then transmitted through premotor networks to modify cortico-cortical connections.

Of the modeling assumptions presented above, the architecture of the network, Hebbian learning with weight normalization, and developmental windows of plasticity are assumptions with a fair amount of experimental support. The assumption involving growth in network gain after premotor circuits have developed is more speculative. We have shown previously that in recurrent networks that learn, low network gain allows the formation of modules that provide a more accurate representation of the set of inputs (S.E. Hua, F.A. Mussa-Ivaldi, and J.C. Houk, in prep.). If network gain is high during learning, reciprocal and modular connections still develop, but internal feedback overshadows the contribution to loop activity by external inputs. However in the present model, PC inhibition may be able to control high levels of feedback activity in the recurrent network.

We propose two major predictions of the model. First, the development of reciprocal and modular connections is a robust feature of the model. Thus, the model predicts that anatomical labeling studies should find results consistent with reciprocal and modular connectivity. We propose that simultaneous labeling of a single motor cortical cell by a retrograde and anterograde viral tracer would allow double labeling of cells involved in the same module in other layers of the network. The second prediction involves electrophysiological evidence for the rotation of the population vector after exportation is complete and is presented in a later section on computational modules.

TOPOGRAPHIC ALIGNMENT OF THE NETWORK

Although we describe the topographic organization of the weights in this study, the weights may be described more accurately in terms of functional topology. Because the layers are of one dimension and wrapped, it is difficult to directly apply the topography of these layers to the three-dimensional brain. More important, the topological organization of the connections come from the correlation structure of the inputs and of the network activity. We allowed the inputs (cortical and PC) to be topographically correlated so that the organization of the connections would be easier to visualize. However, such topographic correlations need not exist. Instead, the correlations used in this model belong to a more general class of topologic or functional correlations. Neighboring units on each layer do not necessarily represent spatial neighbors; instead, neighboring units are functional neighbors. Correlated PC activity represents correlations of directional tuning or muscle synergies and not necessarily spatial topographic relationships. Thus, the actual topographic layout of modularity in the three-dimensional brain is difficult to predict from this study, because the results predict mainly functional topology. Additionally, simple functional topologic relationships such as orientation tuning have complex two-dimensional topography in the visual cortex.

COMPUTATIONAL MODULES FOR ELEMENTAL MOTOR COMMANDS

The present results illustrate that reciprocity is a natural outcome of Hebbian learning in recurrent networks. Strict reciprocity exists when activity initiated in one unit will only return to that one unit without divergence to other units on that layer. In contrast, the present simulation results show that reciprocal connections with divergent recurrent connections develop in the premotor circuit. The divergent connections are made between units having correlated activity such as neighboring cortical units with lateral connections or neighboring cerebellar nuclear units with common PC disinhibition. Whereas reciprocal connections sustain activity in a particular premotor circuit, divergent recurrent connections allow recruitment of premotor circuits with similar functional properties, that is, correlated activity. Highly correlated premotor circuits with strong, divergent interconnections are considered as computational modules that cooperate to produce an elemental motor command (Houk et al. 1993). Weaker divergent connections between different modules serve to combine elemental motor commands under the guidance of the cerebellar cortex.

In the present model, divergent connections allow the population direction vector to rotate from a shifted visual target to the corrected movement direction derived from cerebellar learning (see also Eisenman et al. 1991). Although the visuomotor task simulated in the present model was prism learning, the present results provide similar population behavior as that seen in the rotation of the population direction vector in monkey motor cortex (Lurito et al. 1991). In these experiments, monkeys made arm movements (hand-held manipulandum) to one of eight equally spaced target lights placed in a circle around the starting position on a two-dimensional surface. Interspersed with direct arm movements to dimly lit targets, monkeys were trained to make arm movements 90° counterclockwise (CCW) to a bright target light. Motor cortical single unit recording showed that the population direction vector initially points toward the nonrotated target light and subsequently rotates to the 90° CCW target.

Rotation of the population vector similar to the experiment of Lurito et al. (1991) is seen in the present model when the cerebellum has learned the new visuo-motor mapping but this learning has not yet been exported to the motor cortex (Fig. 6). After cerebellar learning has been exported to the cortex, the population vector no longer rotates because cortico-cortical connections have learned to initiate movement in the correct direction (Fig. 7). These results predict that if monkeys were trained on only the rotated arm movements, rotation of the population vector would diminish as cerebellar learning is exported to cortical sites. Our results further suggest that the presence of the rotating population vector in monkey experiments may occur at a stage where the cerebellum has learned the correct visuo-motor mapping, but this learning has not been exported to the cortex. It is possible that the intermixing of nonrotated and rotated movements in the same block may have slowed the exportation of the rotated visuo-motor task.

The present model demonstrates that the initiation of movements in the appropriate direction can be exported from the cerebellum to cortical sites. This model is a first step in describing the exportation of both movement initiation and execution to motor cortex. We have shown previously that the APG model of cerebellar function programs movement execution by a dynamic pattern of PC activation. In this way, movements are terminated by the activation of PCs. The exportation of movement execution and termination to cortical networks remains a difficult problem and will be approached in future modeling efforts.

Acknowledgments

We thank Ferdinando Mussa-Ivaldi and Andrew Barto for helpful discussions. This work was supported by National Institutes of Health/National Institute of Mental Health grant 5-P50-MH48185 to J.C.H.

The publication costs of this article were defrayed in part by payment of page charges. This article must therefore be hereby marked "advertisement" in accordance with 18 USC section 1734 solely to indicate this fact.

References

Albus, J.S. 1971. A theory of cerebellar function. *Math. Biosci.* **10:** 25–61.

Anderson, J.A., J.W. Silverstein, S.A. Ritz, and R.S. Jones. 1977. Distinctive features, categorical perception, and probability learning: Some applications of a neural model. *Psychol. Rev.* **84:** 413–451.

Bauer, J. and R. Held. 1971. Comparison of visually guided reaching in normal and deprived infant monkeys. *J. Exp. Psychol. Anim. Behav. Processes* **1:** 298–308.

Bear, M.F. and A. Kirkwood. 1993. Neocortical long-term potentiation. *Curr. Opin. Neurobiol.* **3:** 197–202.

Berthier, N.E., S.P. Singh, A.G. Barto, and J.C. Houk. 1993. Distributed representation of limb motor programs in arrays of adjustable pattern generators. *J. Cognit. Neurosci.* **5:** 56–78.

Bloedel, J.R., T.J. Ebner, and S.P. Wise. 1996. *Acquisition of motor behavior in vertebrates.* MIT Press, Cambridge, MA.

Brodal, P. and J.G. Bjaalie. 1992. Organization of the pontine nuclei. *Neurosci. Res.* **13:** 83–118.

Brown, T.H., E.W. Kairiss, and C.L. Keenan. 1990. Hebbian synapses: Biophysical mechanisms and algorithms. *Annu. Rev. Neurosci.* **13:** 475–511.

Crair, M.C. and R.C. Malenka. 1995. A critical period of long-term potentiation at thalamocortical synapses. *Nature* **375:** 325–328.

Ebner, T.J., D. Flament, and S.J. Shanbhag. 1996. The cerebellum's role in voluntary motor learning: Clinical, electrophysiological, and imaging studies. In *Acquisition of motor behavior in vertebrates* (ed. J.R. Bloedel, T.J. Ebner, and S.P. Wise), pp. 235–260. MIT Press, Cambridge, MA.

Eisenman, L.N., J. Keifer, and J.C. Houk. 1991. Positive feedback in the cerebro-cerebellar recurrent network may explain rotation of population vectors. In *Analysis and modeling of neural systems* (ed. F. Eeckman), pp. 371–376. Kluwer Academic Publishers, Norwell, MA.

Fox, K. and K. Zahs. 1994. Critical period control in sensory cortex. *Curr. Opin. Neurobiol.* **4:** 112–119.

Fox, K., B.L. Schlaggar, S. Glazewski, and D.D.M. O'Leary. 1996. Glutamate receptor blockade at cortical synapses disrupts development of thalamocortical and columnar organization in somatosensory cortex. *Proc. Natl. Acad. Sci.* **93:** 5584–5589.

Friston, K.J., C.D. Frith, R.E. Passingham, P.F. Liddle, and R.S. Frackowiak. 1992. Motor practice and neurophysiological adaptation in the cerebellum: A positron tomography study. *Proc. R. Soc. Lond. - Ser. B: Biol. Sci.* **248:** 223–228.

Galiana, H.L. 1986. A new approach to understanding adaptive visual-vestibular interactions in the central nervous system. *J. Neurophysiol.* **55:** 349–374.

Ghosh, S., G.M. Murray, A.B. Turman, and M.J. Rowe. 1994. Corticothalamic influences on transmission of tactile information in the ventro-posterolateral thalamus of the cat: Effect of reversible inactivation of somatosensory cortical areas I and II. *Exp. Brain Res.* **100:** 276–286.

Gilbert, P.F.C. 1974. A theory of memory that explains the function and structure of the cerebellum. *Brain Res.* **70:** 1–18.

Hayes, W.P. and R.L. Meyer. 1988a. Optic synapse number but not density is maintained during regeneration onto surgically-halved tectum in goldfish: HRP-EM evidence that optic fibers compete for fixed numbers of postsynaptic sites on tectum. *J. Comp. Neurol.* **274:** 539–559.

———. 1988b. Retinotopically inappropriate synapses of subnormal density formed by surgically misdirected optic fibers in goldfish tectum. *Brain Res.* **38:** 304–312.

Houk, J.C. 1997. On the role of the cerebellum and basal ganglia in cognitive signal processing. *Prog. Brain Res.* **114:** 545–554.

Houk, J.C. and A.G. Barto. 1992. Distributed sensorimotor learning. In *Tutorials in motor behavior II* (ed. G.E. Stelmach and J. Requin), pp. 71–100. Elsevier, Amsterdam, The Netherlands.

Houk, J.C., S.P. Singh, C. Fisher, and A.G. Barto. 1990. An adaptive sensorimotor network inspired by the anatomy and physiology of the cerebellum. In *Neural networks for control* (ed. W.T. Miller, R.S. Sutton, and P.J. Werbos), Chapter 13, pp. 301–348. MIT Press, Cambridge, MA.

Houk, J.C., J. Keifer, and A.G. Barto. 1993. Distributed motor commands in the limb premotor network. *Trends Neurosci.* **16:** 27–33.

Iriki, A., C. Pavlides, A. Keller, and H. Asanuma. 1991. Long term potentiation of thalamic input to the motor cortex induced by coactivation of thalamocortical and corticocortical afferents. *J. Neurophysiol.* **65:** 1435–1441.

Ito, M. 1989. Long-term depression. *Annu. Rev. Neurosci.* **12:** 85–102.

Keifer, J. 1996. Effects of red nucleus inactivation on burst discharge in turtle cerebellum: Evidence for positive feedback. *J. Neurophysiol.* **76:** 2200–2210.

Keifer, J., D. Vyas, and J.C. Houk. 1992. Sulforhodamine labeling of neural circuits engaged in motor pattern generation in the in vitro turtle brainstem-cerebellum. *J. Neurosci.* **12:** 3187–3199.

Kim, S.-G., K. Ugurbil, and P.L. Strick. 1994. Activation of a cerebellar output nucleus during cognitive processing. *Science* **265:** 949–951.

Lurito, J.T., T. Georgakopoulos, and A.P. Georgopoulos. 1991. Cognitive spatial-motor processes 7. The making of movements at an angle from a stimulus direction: studies of motor cortical activity at the single cell and population levels. *Exp. Brain Res.* **87:** 562–580.

Ma, W. and S.L. Juliano. 1991. The relationship between thalamocortical connections and stimulus-evoked metabolic activity in the ventroposterior nucleus of the monkey. *Somatosens Mot. Res.* **8:** 77–86.

Markram, H. and M. Tsodyks. 1996. Redistribution of synaptic efficacy between neocortical pyramidal neurons. *Nature* **382:** 807–810.

Marr, D. 1969. A theory of cerebellar cortex. *J. Physiol. (Lond.)* **202:** 437–470.

McCormick, D.A. 1989. Cholinergic and noradrenergic modulation of thalamocortical processing. *Trends Neurosci.* **12:** 215–221.

Milak, M.S., V. Bracha, and J.R. Bloedel. 1995. Relationship of simultaneously recorded cerebellar nuclear neuron discharge to the acquisition of a complex, operantly conditioned forelimb movement in cats. *Exp. Brain Res.* **105:** 325–330.

Miller, K.D. 1994. A model for the development of simple cell receptive fields and the ordered arrangement of orientation columns through activity-dependent competition between ON- and OFF-center inputs. *J. Neurosci.* **14:** 409–441.

Miller, K.D. and D.J.C. MacKay. 1994. The role of constraints in Hebbian learning. *Neural Comput.* **6:** 100–126.

Miller, K.D., J.B. Keller, and M.P. Stryker. 1989. Ocular dominance column development: Analysis and simulation. *Science* **245:** 605–615.

Petralia, R.S., N. Yokotani, and R.J. Wenthold. 1994. Light and electron microscope distribution of the NMDA receptor subunit NMDAR1 in the rat nervous system using a selective anti-peptide antibody. *J. Neurosci.* **14:** 667–696.

Prince, D.A. and J.R. Huguenard. 1988. Functional properties of neocortical neurons. In *Neurobiology of neocortex* (ed. P. Rakicand and W. Singer), pp. 153–176. John Wiley & Sons, Chichester, UK.

Raichle, M.E., J.A. Fiez, T.O. Videen, A.K. Macleod, J.V. Pardo, P.T. Fox, and S.E. Petersen. 1994. Practice-related changes in human brain functional anatomy during nonmotor learning. *Cereb. Cortex* **4:** 8–26.

Raymond, J.L., S.G. Lisberger, and M.D. Mauk. 1996. The cerebellum: A neuronal learning machine? *Science* **272:** 1126–1131.

Shinomoto, S. 1987. Memory maintenance in neural networks. *J. Phys. A: Math. Gen.* **20:** L1305–L1309.

Thach, W.T., H.P. Goodkin, and J.G. Keating. 1992. Cerebellum and the adaptive coordination of movement. *Annu. Rev. Neurosci.* **15:** 403–442.

Thompson, R.F. 1986. The neurobiology of learning and memory. *Science* **233:** 941–947.

Tsukahara, N., N. Bando, T. Murakami, and Y. Oda. 1983. Properties of cerebello-precerebellar reverberating circuits. *Brain Res.* **274:** 249–259.

Tyrrell, T. and D. Willshaw. 1992. Cerebellar cortex: Its simulation and the relevance of Marr's theory. *Philos. Trans. R. Soc. Lond. - Ser. B: Biol. Sci.* **336:** 239–257.

Whitelaw, V.A. and J.D. Cowan. 1981. Specificity and plasticity of retinotectal connections: A computational model. *J. Neurosci.* **1:** 1369–1387.

Zhong, J., D.P. Carrozza, K. Williams, D.B. Pritchett, and P.B. Molinoff. 1995. Expression of mRNAs encoding subunits of the NMDA receptor in developing rat brain. *J. Neurochem.* **64:** 531–539.

Received February 24, 1997; accepted in revised form May 7, 1997.

Role of Cerebellum in Adaptive Modification of Reflex Blinks

John J. Pellegrini[1] and Craig Evinger[2,3]

[1]Department of Biology
The College of St. Catherine
St. Paul, Minnesota 55105
[2]Departments of Neurobiology and Behavior and Ophthalmology
State University of New York at Stony Brook
Stony Brook, New York 11794-5230

Abstract

We investigated the involvement of the cerebellar cortex in the adaptive modification of corneal reflex blinks and the regulation of normal trigeminal reflex blinks in rats. The ansiform Crus I region contained blink-related Purkinje cells that exhibited a complex spike 20.4 msec after a corneal stimulus and a burst of simple spike activity correlated with the termination of orbicularis oculi activity. This occurrence of the complex spike correlated with trigeminal sensory information associated with the blink-evoking stimulus, and the burst of simple spike activity correlated with sensory feedback about the occurrence of a blink. Inactivation of the inferior olive with lidocaine prevented all complex and significantly reduced simple spike modulation of blink-related Purkinje cells, but did not alter orbicularis oculi activity evoked by corneal stimulation. In contrast, both acute and chronic lesions of the cerebellar cortex containing blink-related Purkinje cells blocked adaptive increases in orbicularis oculi activity of the lid ipsilateral but not contralateral to the lesion. These data are consistent with the hypothesis that the cerebellum is part of a trigeminal reflex blink circuit. Changes in trigeminal signals produce modifications of the cerebellar cortex, which in turn, reinforce or stabilize modifications of brainstem blink circuits. When the trigeminal system does not attempt to alter the magnitude of trigeminal reflex blinks, cerebellar input has little or no effect on reflex blinks.

Introduction

Reflex blinks are a simple motor behavior that undergo rapid adaptive modification in response to altered sensory feedback (Evinger and Manning 1988; Evinger et al. 1989; Schicatano et al. 1996). For example, reducing the motility of the upper eyelid initiates a rapid increase in the magnitude of the lid closing, orbicularis oculi (OO) muscle activity to compensate for the reduction in lid mobility. This adaptive behavior is a learned response rather than a proprioceptive reaction to restriction of lid motion, because the larger than normal OO activity elicited by a blink-evoking stimulus remains for a period *after* returning lid motility to normal. Blink adaptation can also reduce the drive on blink circuits. Assisting lid closure causes a learned reduction in OO activity in response to a blink-evoking stimulus. Therefore, the nervous system must constantly compare the neural representation of the motor command for a blink with the proprioceptive feedback from the actual reflex blink. A consistent difference between these two values initiates an active modification of the trigeminal drive on OO motoneurons to maintain a constant relationship between stimulus and blink magnitude. Recent evidence indicates that these adaptive modifications occur in the spinal trigeminal complex (Schicatano et al. 1996).

Adaptive blink mechanisms can have a significant role in disease. For example, Bell's palsy, resulting from unilateral facial nerve damage, induces changes in trigeminal blink circuits identical to those generated by upper eyelid restraint (Huffman et al. 1996; Schicatano et al. 1996). Such adap-

[3]Corresponding author.

tive modifications assist in compensating for the reduction in lid motility caused by facial nerve damage. Modifications of trigeminal blink circuits initiated by adaptive processes can also cause disease states. Under conditions of decreased inhibition of the trigeminal system, a normally adaptive increase in trigeminal reflex blink excitability can lead to uncontrollable spasms of lid closure, or blepharospasm (Chuke et al. 1996; Schicatano et al. 1997). Therefore, investigating the neural mechanisms underlying adaptive modification of reflex blinks is important for understanding disease processes, as well as for understanding basic mechanisms underlying procedural learning.

The importance of the cerebellum in adaptive modification of the vestibulo-ocular reflex (for review, see duLac et al. 1995) and in conditioned learning of the blink reflex (for review, see Thompson and Krupa 1994) suggests that the cerebellum has a role in adaptive modification of trigeminal reflex blinks. One can envision at least two roles for the cerebellum in adaptive modification of the reflex blinks. Using trigeminal mossy and climbing fiber inputs, the cerebellum could detect errors between the expected and actual blink magnitude. The cerebellum would then actively regulate the magnitude of trigeminal reflex blinks to eliminate discrepancies between an efferent copy signal of the blink and the actual blink. In this scheme, the cerebellum is a regulatory element acting on an autonomous reflex blink circuit. A second possibility is that the cerebellum is an element of the reflex blink circuit. In this scheme, the cerebellum receives trigeminal sensory information and uses this information to reinforce and stabilize modifications initiated within trigeminal circuits. In contrast to the first scheme, the interactions between the cerebellum and the trigeminal system are reciprocal in this second model. The current study reports on the discharge characteristics of blink-related Purkinje cells and explores their role in normal reflex blinks and reflex blink adaptation. The data indicate that the trigeminal system and the cerebellum exert reciprocal effects when the nervous system adapts the gain-of-trigeminal reflex blinks.

Materials and Methods

PROCEDURES

Subjects for all experiments were male Sprague-Dawley rats. The details of the procedures used in this study have been reported in previous studies (Evinger et al. 1993; Basso et al. 1996). The animals were prepared for recording electromyograph (EMG) activity of the lid closing, orbicularis oculi muscle (OOemg) and single-unit, extracellular recording. Two preparations were employed. In initial experiments, rats were anesthetized with ketamine (90 mg/kg) and xylazine (10 mg/kg) and then underwent precollicular decerebration by aspiration. No further anesthesia was administered. In later experiments, rats were anesthetized with urethane [15 grams/kg in phosphate buffer, pH 7.4, intraperitoneally (i.p.)] and no decerebration was performed. To record OOemg, a pair of single-stranded, Teflon-coated, stainless steel wires were implanted into the medial and lateral margins of both the right and left eyelids. To allow restraint of the eyelid, a silk suture was attached to the center of the lower margin of both upper eyelids. To provide access to the cerebellum for microelectrodes, the bone overlying the Crus I region of the cerebellum was removed. In some experiments, a hole was drilled in the skull overlying the inferior olive to allow access to this region with microelectrodes. In other experiments, either an acute or chronic unilateral lesion of the Crus I region was performed by aspiration. For chronic lesions, the rat received general anesthesia (ketamine, 90 mg/kg; xylazine, 10 mg/kg). Using aseptic techniques, the skull overlying the left Crus I region was removed and the underlying cerebellar cortex was aspirated. The bone defect was covered with bone wax and the overlying skin sutured. Rats were allowed to recover from anesthesia and treated with analgesics as necessary. Three or four days later, rats were reanesthetized and prepared as described above. For acute lesions, an aspiration of Crus I cerebellar cortex was performed during the experiment. At the end of all experiments, the rats were perfused with 6% dextran to wash out blood, followed by 10% formalin to allow histological examination of lesions or location of fast green dye injections.

Blinks were evoked by electrical stimulation of the cornea through a pair of silver ball electrodes. In most cases, a single 100-μsec stimulus elicited a corneal reflex blink. In other cases, it was necessary to present three 100-μsec stimuli to evoke a corneal reflex blink. To test the excitability of corneal reflex blinks, pairs of identical stimuli with a 75-msec interstimulus interval were presented, the paired stimulus paradigm (Pellegrini and Evinger 1995). To avoid habituation, the interval between stimuli (or pairs of stimuli) was 45 ± 5 sec.

To induce adaptive increases in the drive to orbicularis oculi motoneurons, the upper eyelids were restrained by connecting the suture attached to the upper eyelid to a fixed bar. Identical corneal stimuli were presented before and after lid restraint. Both the left and right cornea received stimuli at the same time.

Single Purkinje cells and interpositus neurons were recorded extracellularly with glass micropipettes filled with 2 M NaCl saturated with fast green. The signals were filtered from 1 to 10 kHz. Purkinje cells were identified by their characteristic complex spike activity. To anesthetize the inferior olive, we used double-barreled electrodes. One barrel was filled with 2% lidocaine and the other barrel was filled with 2 M NaCl saturated with fast green. The lidocaine barrel was attached to a Picospritzer (General Valve Co.) to allow brief microinjections of lidocaine. To determine the efficacy of inferior olive inactivation, we simultaneously recorded from blink-related Purkinje cells during corneal stimulation. We considered the inferior olive inactivated, when corneal stimulation no longer elicited complex spikes in Purkinje cells, but the OOemg activity of corneal-evoked blinks was still present. Fast green dye injections (Thomas and Wilson 1965) were made at the site of Purkinje cell recordings and lidocaine injections.

DATA COLLECTION AND ANALYSIS

OOemg, single-unit activity, and the stimuus were stored on analog tape for later off-line analysis. OOemg data were digitized (5 kHz per channel, 12-bit precision) and analyzed using laboratory-developed software. The OOemg records were rectified and integrated and averaged to determine OOemg magnitude. Individual simple spikes were recognized with a window discriminator and the output of the discriminator was digitized (2 kHz, 12-bit precision). Simple spike analysis used raster displays, histograms (10-msec bins), or spike density profiles. The spike density function is a low-pass-filtered gauge of the probability of spike occurrence over time. Our analysis convolved the spike train with Gaussian kernel, σ = 2 msec (for details, see Richmond et al. 1987).

Results

BLINK-RELATED PURKINJE CELLS

Purkinje cells whose discharge correlated with corneal stimulation and blinking were found in specific regions of the cerebellar cortex. Blink-related Purkinje cells were found consistently in Crus I of the ansiform lobule (Fig. 1). No systematic search for all possible locations of blink-related Purkinje cells was performed, but blink-related Purkinje cells responding to corneal stimulation were not found in nearby regions.

In response to stimulation of the cornea, blink-

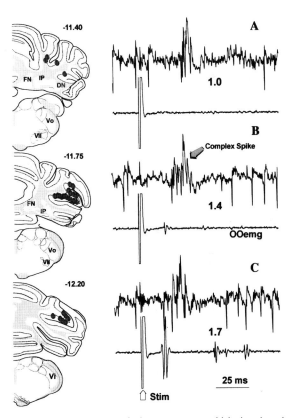

Figure 1: Location and characteristics of blink-related Purkinje cells. (*Left*) Three schematic sections of rat cerebellum adapted from Swanson (1992) located 11.4, 11.75, and 12.2 mm posterior to bregma. Circles show location of blink-related Purkinje cells marked by fast green injections. (*A–C*) Individual records of Purkinje cell (*top* trace) and simultaneous orbicularis oculi EMG activity (*bottom* trace, OOemg) collected at threshold for evoking a complex spike response 1.0 (*A*), 1.4 (*B*), and 1.7 (*C*) times threshold. (DN) Dentate nucleus; (FN) fastigial nucleus; (IP) interpositus nucleus; (Vi) spinal trigeminal nucleus interpolaris subdivision; (Vo) spinal trigeminal nucleus oralis subdivision; (VII) facial nucleus.

related Purkinje cells exhibited a complex spike, followed by a brief pause in simple spike activity and a subsequent burst of simple spike activity (Figs. 1 and 2). The complex spike was the most sensitive component of this evoked response. The complex spike could be elicited by corneal stimuli that were too weak to evoke OOemg activity (Fig. 1, cf. A and C). When the corneal stimulation failed to evoke OOemg activity, the pause in simple spike activity was often longer than when a blink occurred. This increased pause duration probably resulted from the absence of a burst of simple spike activity produced by mossy fiber inputs related to the occurrence of a blink (see below). Following a corneal stimulus that evoked a blink, the complex spike occurred after the onset of OOemg activity. The mean latency of complex spike activity was 20.4 msec (± 3.78, $n = 47$ cells), whereas the mean latency of OOemg activity was only 16.46 msec (± 5.8 msec, $n = 68$ experiments). The burst in simple spike activity typically occurred near or coincident with the end of the OOemg activity and lasted 50–75 msec (Fig. 2). The mean simple spike frequency for a 50-msec period of the burst divided by the mean simple spike frequency during a 50-msec period of spontaneous activity averaged 1.46 (± 0.205, $n = 18$). The mean, maximum firing frequency during the burst was 122 spikes/sec ($n = 47$) for all blink-related Purkinje cells. The burst in Purkinje cell simple spike activity with corneal evoked blinks did not result from a rebound excitation following the complex spike. There was always a burst of simple spike discharge following a complex spike evoked by corneal stimulation that elicited a blink, but no increase of simple spike discharge following spontaneous complex spikes (Fig. 3). Therefore, the burst of simple spike discharge with the termination of OOemg activity carried information about the occurrence of a corneally evoked reflex blink. Because the cerebellum receives mossy fiber inputs from the trigeminal complex (vanHam and Yeo 1992),

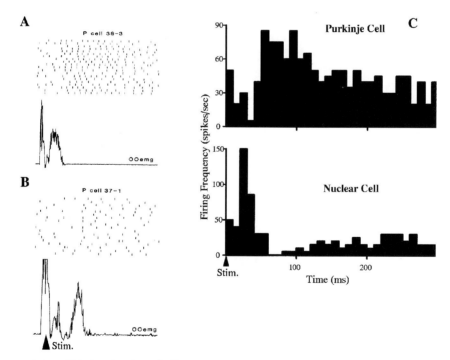

Figure 2: Characteristics of blink-related cerebellar neurons. (*A*) Raster of simple spike activity of a single blink-related Purkinje cell (39–3) and rectified and averaged OOemg activity of 20 consecutive trials of corneal stimulation. Each mark represents the occurrence of a simple spike and each line of marks is a different trial. The duration of the entire record is 250 msec. (*B*) Raster of simple spike activity of a single blink-related Purkinje cell (37–1) and rectified and averaged OOemg activity of 18 consecutive trials with corneal stimulation. Each mark represents the occurrence of a simple spike and each line of marks is a different trial. The duration of the entire record is 150 msec. (*C*) Histograms of simple spike activity of a blink-related Purkinje cell (*top* record) and spikes of a blink-related interpositus nucleus neuron (*bottom* record). Each histogram is the sum of 20 trials using 10-msec bins. The occurrence of the corneal stimulus coincides with the onset of the histogram.

tude was an insignificant 2.9% decrease following inferior olive inactivation (*t*-test, $P > 0.25$; $n = 23$ experiments). Therefore, reducing or eliminating the modulation of blink-related Purkinje cell activity evoked by corneal stimulation did not significantly alter OOemg magnitude or duration.

ADAPTIVE MODIFICATION OF THE BLINK REFLEX

A reduction in eyelid motility induces an increase in the magnitude and excitability of trigeminal reflex blinks (Evinger and Manning 1988; Evinger et al. 1989; Schicatano et al. 1996). These adaptive increases occur almost exclusively in the long latency components of OOemg activity (Evinger and Manning 1988; Evinger et al. 1989). Because this is the only segment of reflex blink activity susceptible to cerebellar influence, the cerebellum could participate in the adaptive increase in the magnitude and duration of reflex blinking. We investigated this possibility by recording the activity of seven blink-related Purkinje cells in six rats before and during lid restraint. In addition, we tested the ability of rats to produce adaptive increases of the blink reflex following cerebellar cortex lesions.

The activity of blink-related Purkinje cells correlated with OOemg activity during lid restraint (Fig. 6). In three of the seven experiments, lid restraint failed to initiate significant increases (0%–4%) in reflex blink magnitude. The Purkinje cells recorded in these experiments did not exhibit significant changes in simple spike activity (Fig. 6B). In the remaining four experiments, the magnitude of OOemg activity increased from 33.5% to 112% over prerestraint values. This elevation of OOemg magnitude occurred primarily in the longer latency components of reflex blink activity and from an increase in blink duration (Fig. 6A). The blink-related Purkinje cells recorded in these experiments all exhibited a backward shift in the time of the occurrence of maximum spike density (Fig. 6C). This delay in the peak simple spike burst should have lengthened the period of interpositus neuron activity associated with the reflex blink (Fig. 2). In contrast to the shift in the timing of peak burst, the tonic simple spike firing frequency in the interblink interval did not change in the brief periods (<40 min) that we tested lid restraint.

Another test of the role of the cerebellar cortex in adaptive modification of the blink reflex was to investigate whether adaptive blink modification occurred following a unilateral lesion of the cerebellar cortex containing blink-related Purkinje cells (Fig. 7). Because the corneal reflex blink is normally unilateral in rats (Basso et al. 1996), it is possible to assess lesion affects on the two lids independently. Three decerebrate rats were challenged with lid restraint that enhanced the amplitude and duration of OOemg activity in response to a corneal stimulus (Fig. 7B, Pre). Restraint was re-

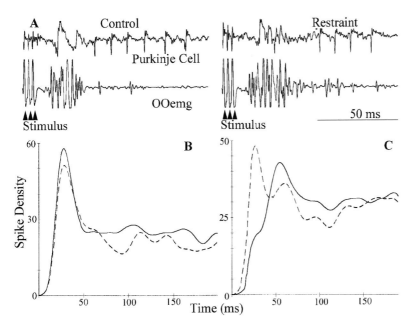

Figure 6: Effect of restraining the upper eyelid on Purkinje cell activity and OOemg responses. (*A*) Individual records of blink-related Purkinje cell and OOemg activity in response to a corneal stimulus (arrowheads, Stimulus) before (Control) and after restraint of the upper eyelid (Restraint). (*B*) Spike density of simple spike activity of a single Purkinje cell before (dashed line) and during (solid line) restraint of the eye lid that did not produce an adaptive increase in OOemg activity. Each spike density record is the average of 20 trials in which $\sigma = 2$ msec. (*C*) Spike density of simple spike activity of a single Purkinje cell before (dashed line) and during (solid line) restraint of the eye lid that produced an adaptive increase in OOemg activity. Each spike density record is the average of 20 trials in which $\sigma = 2$ msec.

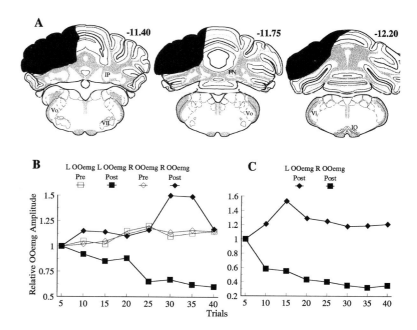

Figure 7: Effect of lesions of the left cerebellar cortex on adaptation of the corneal blink reflex. (A) Reconstruction of a lesion that abolished adaptive gain increases for the left eyelid. Three schematic sections of rat cerebellum adapted from Swanson (1992) 11.4, 11.75, and 12.2 mm posterior to bregma. Solid areas show site of lesion. (B) OOemg amplitude relative to prerestraint OOemg values for the left (□, ■) and right (◇, ◆) OOemg during restraint of the eyelids before (□, ◇) and after (■, ◆) a lesion of the cerebellar cortex depicted in A. Each point is the average of five trials. (C) OOemg amplitude relative to prerestraint OOemg values for the left (■) and right (◆) OOemg during restraint of the upper eyelids three days after a lesion of the left cerebellar cortex. Each point is the average of five trials.

moved and OOemg activity returned to prerestraint levels (not illustrated). Aspiration in the left cerebellar cortex region containing blink-related Purkinje cells, altered the response to lid restraint in two of the rats. The right OOemg exhibited normal increases in OOemg activity, whereas the left OOemg exhibited a decrease in activity (Fig. 7B, Post). Histology revealed that the lesions in these two animals completely ablated cortical regions containing blink-related Purkinje cells (Fig. 7A). The third rat, with an incomplete lesion of the cerebellar cortex containing blink-related Purkinje cells, exhibited increases in OOemg activity nearly identical to that seen before the lesions. In another group of three rats, we performed lesions in the left cerebellar cortex three or four days before decerebration and testing of the adaptive modification of the corneal reflex blink. In the two animals where the earlier lesion completely abolished the cortical areas containing blink-related Purkinje cells, the left OOemg failed to adapt, whereas the right OOemg exhibited normal adaptation (Fig. 7C).

Discussion

ROLE OF THE CEREBELLUM IN TRIGEMINAL REFLEX BLINKS

Blink-related Purkinje cell activity can only act on the longer latency components of OOemg activity evoked by a corneal stimulus. The first Purkinje cell response to corneal stimulation, a complex spike, occurs after the initiation of OOemg activity. Likewise, the burst of spikes in rat deep cerebellar neurons is also too late to initiate OOemg activity because the burst appears to take place at the same time as the complex spike in Purkinje cells. A study of deep cerebellar nucleus neurons in cats reached an identical conclusion (Gruart and Delgado-García 1994). The burst of simple spike activity of blink-related Purkinje cells and the pause in interpositus neuron activity correlates with the termination of OOemg activity. Therefore, the cerebellum could modulate the long latency components and the duration of trigeminal reflex blinks, but the cerebellum could not influence the initiation of corneal reflex blinks.

A possible flaw in the previous arguments about the ability of the cerebellum to modulate the initiation of trigeminal reflex blinks is that changes in the tonic level of interpositus activity could provide moment-to-moment regulation of all reflex blink components. The results of inferior olive inactivation, however, argue against this possibility. As reported previously (Benedetti et al. 1983; Batini et al. 1985), inactivation of the inferior olive increases the level of tonic Purkinje cell simple spike activity (Fig. 5). This increased simple spike activity causes a decrease in deep cerebellar neuron activity and excitability. These changes in tonic activity, however, fail to modify reflex blinks. Therefore, altering cerebellar activity fails to modify trigeminal reflex blinks in the short term. Previous studies of the effect of cerebellar damage on reflex blinking are conflicting. Some investigators report that cerebellar damage does not affect unconditioned blink responses (for review, see Steinmetz et al. 1992), although others find a decrement in unconditioned response magnitude (for review, see Bloedel and Bracha 1995). Yeo and Hardiman (1992) report an increase in the magnitude of unconditioned responses following cerebellar cortical lesions.

The activity of blink-related Purkinje cells appears to reflect sensory- rather than motor-related signals. The complex spike evoked in blink-related Purkinje cells is a trigeminal sensory signal, as the complex spike can be elicited in the absence of OOemg activity (Fig. 1). Likewise, the burst of simple spike activity correlated with the end of OOemg activity is absent when blinks are reduced in amplitude as occurs in the paired stimulus paradigm (Fig. 4). The reduction in the burst of simple spike activity following inactivation of the inferior olive (Fig. 5) suggests that the occurrence of a simple spike acts to gate mossy fiber trigeminal sensory signals. The paired stimulus paradigm also demonstrates that suppression occurring within the spinal trigeminal complex (Pellegrini and Evinger 1995) can eliminate evoked complex spikes in blink-related Purkinje cells as well as the burst of simple spike activity.[1] If, as the current data indicate, the blink-related Purkinje cells exhibit a sensory rather than a motor signal, it is not surprising that eliminating Purkinje cell modulation fails to modify trigeminal reflex blinks in the short term (Fig. 5).

ROLE OF THE CEREBELLUM IN ADAPTIVE MODIFICATION OF TRIGEMENAL REFLEX BLINKS

Despite the evidence against a role for the cerebellum in regulating reflex blinks, the current lesion and recording data suggest a powerful role for the cerebellum in adaptive modification of trigeminal reflex blinks. This incongruity can be reconciled by assuming that trigeminal sensory signals modify cerebellar circuits which, in turn, reinforce or stabilize desired synaptic changes within trigeminal reflex blink pathways. In this scheme, reorganization of trigeminal signals alter cerebellar activity. Consistent changes in trigeminal signals initiated by adaptive modification may "teach" the cerebellum a new firing pattern. This new firing pattern reinforces or stabilizes the adaptive modifications in trigeminal reflex blink circuits. Without the cerebellum, attempts to alter the gain-of-trigeminal blink reflexes fail to develop appropriately. Nevertheless, although there is no drive to alter trigeminal reflex blinks, removal of the cerebellum will not effect trigeminal reflex blinks in the short term. After prolonged period of cerebellar damage, however, we would expect the gain-of-trigeminal reflex blinks to drift. This suggestion is consistent with the conflicting reports about the effect of cerebellar damage on reflex blinks in conditioned eyelid studies.

Aspiration of the cerebellar cortex containing blink-related Purkinje cells not only prevents adaptive increases in OOemg magnitude, cerebellar cortex damage leads to a decrease in OOemg magnitude with lid restraint (Fig. 7). A similar decrease in blink amplitude following lid restraint occurs after a lesion of the supraorbital branch of the trigeminal nerve which provides sensory feedback from the blink (Evinger et al. 1989). If trigeminal sensory inputs modulate the activity of blink-related Purkinje cells, it makes sense that a loss of trigeminal sensory input and elimination of blink-related Purkinje cells produce the same disruption of adaptive gain modification. The reason that blink magnitude decreases in these two conditions may be the relative weightings of inhibitory and facilitatory forces that sum to produce a reflex blink. A trigeminal blink-evoking stimulus probably initiates both a positive feedback circuit to produce the blink and a slower acting inhibitory process to terminate the blink (Evinger 1995). An adaptive increase in blink magnitude can occur by delaying or reducing this inhibitory process. If so, attempts to modify blink magnitude destabilize the inhibitory processes. When changes in the cerebellar output do not reinforce rearrangements of the inhibitory processes, they tend to drift toward higher levels, which would reduce blink magnitude. Because uncontrollable spasms of lid closure result if the inhibitory processes drift toward lower levels (Schicatano et al. 1997), the system probably has a built in tendancy to move toward more rather than less inhibition when adaptation occurs.

---

[1] Because all blinks induce a transient suppression of trigeminal inputs (Powers et al. 1997), the absence of a complex spike evoked by an unconditioned stimulus following a conditioned response may be attributable to trigeminal inhibition rather than interpositus inhibition of the inferior olive.

Although the current study primarily examined the cerebellar cortex, the results do not imply that the cerebellar cortex is the sole site of plasticity associated with adaptive gain changes. A recent model of the cerebellum (Mauk and Donegan, this issue) illustrates how cerebellar plasticity can be distributed between the cerebellar cortex and the cerebellar nuclei. In the initial stages of plasticity, the cerebellar cortex is crucial in teaching the deep cerebellar nuclei. Therefore, lesions of the cerebellar cortex in the initial stages of gain changes would disrupt adaptive plasticity. After adaptation had occurred, cortical lesions would exert little effect on reflex blinks, because the deep cerebellar nuclei had already undergone gain modification. The current data are completely consistent this model.

The data from the current and a previous (E.J. Schicatano, K.R. Peshori, V.M. Henriquez, and C. Evinger, unpubl.) study suggest that the trigeminal complex, rather than the cerebellum, is the pivotal site for adaptive modification of the trigeminal reflex blinks. Many studies of Pavlovian conditioning of the eye blink argue that the cerebellum is the pivotal site for acquisition of conditioned responses (e.g., Thompson and Krupa 1994). Both of these views are consistent with the current scheme of cerebellar regulation of blinking. The sensory cues necessary for adaptive modification of reflex blink magnitude, motor efference copy, and sensory feedback from the movement are both immediately available in the brainstem trigeminal complex. In contrast, the association of an auditory signal and a reflex blink necessary for the creation of a conditioned response are more readily available in the cerebellum than in the trigeminal complex. Therefore, the two paradigms rely more heavily on one portion of the circuit rather than the other. If the cerebellum is an integral component in a trigeminal reflex blink circuit, then changes in either the trigeminal system or in the cerebellum will modify other components of the circuit. The current data show how modifications occurring within the trigeminal system alter the response of blink-related Purkinje cells, as in the paired stimulus paradigm (Fig. 4). Likewise, the acquisition of conditioned responses alters the threshold and excitability of the trigeminal complex (E. Schicatano and C. Evinger, unpubl.). Therefore, the simplest interpretation of these data is that the interactions between the cerebellum and the trigeminal system are reciprocal and that both undergo plastic changes that reinforce each other. In adaptive changes, the trigeminal complex is the driving force, whereas in eye-blink conditioning, the cerebellum is the driving force.

Acknowledgments

We thank Donna Schmidt for her excellent technical assistance. This work was supported by grant EY07391 from the National Eye Institute (C.E.).

The publication costs of this article were defrayed in part by payment of page charges. This article must therefore be hereby marked ''advertisement'' in accordance with 18 USC section 1734 solely to indicate this fact.

References

Basso, M.A., A.S. Powers, and C. Evinger. 1996. An explanation for blink reflex hyperexcitability in Parkinson's disease. I. Superior colliculus. *J. Neurophysiol.* **16:** 7308–7317.

Batini, C., J.M. Billard, and H. Daniel. 1985. Long-term modification of cerebellar inhibition after inferior olive degeneration. *Exp. Brain Res.* **59:** 404–409.

Benedetti, F., P.G. Montarolo, P. Strata, and F. Tempia. 1983. Inferior olive inactivation decreases the excitability of the intracerebellar and lateral nuclei in the rat. *J. Physiol. (Lond.)* **340:** 195–208.

Bloedel, J.R. and V. Bracha. 1995. On the cerebellum, cutaneomuscular reflexes, movement control and the elusive engrams of memory. *Behav. Brain Res.* **68:** 1–44.

Chuke, J.C., R.S. Baker, and J.D. Porter. 1996. Bell's palsy—associated blepharospasm relieved by aiding eye closure. *Ann. Neurol.* **39:** 263–268.

duLac, S., J.L. Raymond, T.J. Sejenowski, and S.G. Lisberger. 1995. Learning and memory in the vestibulo-ocular reflex. *Annu. Rev. Neurosci.* **18:** 409–442.

Evinger, C. 1995. A brainstem reflex in the blink of an eye. *News Physiol. Sci.* **10:** 147–153.

Evinger, C. and K.A. Manning. 1988. A model system for motor learning: Adaptive gain control of the blink reflex. *Exp. Brain Res.* **70:** 527–538.

Evinger, C., J.J. Pellegrini, and K.A. Manning. 1989. Adaptive gain modification of the blink reflex. A model system for investigating the physiologic bases of motor learning. *Annu. N.Y. Acad. Sci.* **253:** 87–100.

Evinger, C., M.A. Basso, K.A. Manning, P.A. Sibony, J.J. Pellegrini, and A.K.E. Horn. 1993. A role for the basal ganglia in nicotinic modulation of the blink reflex. *Exp. Brain Res.* **92:** 507–515.

Gruart, A. and J.M. Delgado-García. 1994. Discharge of identified deep cerebellar nuclei neurons related to eye blinks in the alert cat. *Neuroscience* **61:** 665–681.

Huffman, M.D., R.S. Baker, M.W. Stava, J.C. Chuke, B.R. Rouholiman, and J.D. Porter. 1996. Kinematic analysis of eyelid movements in patients recovering from unilateral facial nerve palsy. *Neurology* **39:** 263–268.

Mauk, M.D. and N.H. Donegan. 1997. A model of Pavlovian eyelid conditioning based on the synaptic organization of the cerebellum. *Learn. & Mem.* (this issue).

Pellegrini, J.J. and C. Evinger. 1995. The trigeminally evoked blink reflex: II. Mechanisms of paired stimulus suppression. *Exp. Brain Res.* **107:** 181–196.

Pellegrini, J.J., A.K.E. Horn, and C. Evinger. 1995. The trigeminally evoked blink reflex: I. Neuronal circuits. *Exp. Brain Res.* **107:** 166–180.

Powers, A.S., E.S. Schicatano, M.A. Basso, and C. Evinger. 1997. To blink or not to blink: Inhibition and facilitation of reflex blinks. *Exp. Brain Res.* **113:** 283–290.

Richmond, B.J., L.M. Optican, M. Podell, and H. Spitzer. 1987. Temporal encoding of two-dimensional patterns by single units in primate primary visual cortex. I. Stimulus-response relationships. *J. Neurophysiol.* **57:** 132–146.

Schicatano, E.J., M.A. Basso, and C. Evinger. 1997. An animal model explains the origins of the cranial dystonia benign essential blepharospasm. *J. Neurophysiol.* (in press).

Steinmetz, J.E., D.G. Lavond, D. Ivkovich, C.G. Logan, and R.F. Thompson. 1992. Disruption of classical eyelid conditioning after cerebellar lesions: Damage to a memory trace system or a simple performance deficit? *J. Neurosci.* **12:** 4403–4426.

Swanson, L.W. 1992. *Structure of the rat brain.* Elsevier, Amsterdam, The Netherlands.

Thomas, R.C. and V.J. Wilson. 1965. Precise localization of Renshaw cells with a new marking technique. *Nature* **206:** 211–213.

Thompson, R.F. and D.J. Krupa. 1994. Organization of memory traces in the mammalian brain. *Annu. Rev. Neurosci.* **17:** 519–549.

van Ham, J.J. and C.H. Yeo. 1992. Somatosensory trigeminal projections to the inferior olive, cerebellum and other precerebellar nuclei in rabbits. *Eur. J. Neurosci.* **4:** 302–317.

Yeo, C.H. and M.J. Hardiman. 1992. Cerebellar cortex and eye blink conditioning: A reexamination. *Exp. Brain Res.* **88:** 623–638.

Received February 18, 1997; revised version accepted April 30, 1997.

Single-Unit Evidence for Eye-Blink Conditioning in Cerebellar Cortex is Altered, but Not Eliminated, by Interpositus Nucleus Lesions

Donald B. Katz and Joseph E. Steinmetz[1]

Department of Psychology and Program in Neural Science
Indiana University
Bloomington, Indiana 47405

Abstract

Many theories of motor learning explain learning-related changes in motor behavior in terms of plasticity in the cerebellar cortex. Empirical evidence, however, does not always appear to be consistent with such formulations. It is the anterior cerebellar interpositus nucleus (aINP) that seems to be essential for acquisition and retention of conditioned eye-blink responses under most circumstances and it has been therefore suggested that the aINP is the critical site of learning-related plasticity during eye-blink conditioning. Supporting this conclusion are studies demonstrating that multiple-unit conditioning-related neural activity patterns observed in many brain regions disappear after aINP lesion. The possibility that the cerebellar cortex may be involved in forming these patterns has not been assessed adequately, however.

In the current study, trained rabbits received kainic acid lesions of the INP. After recovery, the animals underwent additional sessions of conditioning during which single-unit activity was recorded from the cerebellar cortex. Our results suggest that the aINP is not the sole site of plasticity during eye-blink conditioning, as a subset of the neurons recorded from lesioned animals demonstrated conditioning-related firing patterns. The lesions did change the character of these firing patterns from those observed in saline controls, however, in ways that can be generally described as a loss of organization. The normal tendency for the population of cortical cells to change firing rate together, for instance, was significantly less noticeable in lesioned animals. These results suggest that the aINP may be involved in the production of important features of conditioned responding, such as system timing function, therefore suggesting the need for more models that incorporate aINP and brain stem feedback as integral to the production of organized neural and behavioral responses.

Introduction

In 1969, David Marr presented a hypothesis of motor learning whereby sensory or cerebrally guided movement commands are associated with teaching or error signals within the cerebellum (Marr 1969). This hypothesis, later modified by James Albus (1971), invokes circuitry intrinsic to cerebellar cortex to explain changes in motor responses in relation to environmental contingencies. The mossy/parallel fiber system is postulated to code movement-related signals, and climbing fibers are postulated to code teaching and error signals by which the success of movement is evaluated. According to the Marr–Albus theory, the result of covariance between these two signals is a loss of efficacy at Purkinje cell–parallel fiber synapses; this plasticity is presumed to code learned movement.

The appeal of the Marr–Albus formulation is such that many, if not most, of the later proposed

[1]Corresponding author.

models of cerebellar function are in some way instantiations of the original theory (for review, see Chapeau-Blondeau and Chauvet 1991; Kawato and Gomi 1992; Ito 1993). All of the cited references postulate that long-term changes in synaptic efficacy develop somewhere in the mossy fiber-Purkinje cell system, and that it is in particular the mossy fiber changes in response to covariant climbing fiber input that underlies motor learning. Most models of eye-blink conditioning rely similarly on the architecture of the cerebellar cortex (and its associated input pathways) to explain the development of an association between a tone conditioned stimulus (CS) and an air puff unconditioned stimulus (US), and to explain the well-timed quality of the resultant conditioned eye blink (Moore and Blazis 1989; Buonomano and Mauk 1994). These models share the assumption that the occurrence of climbing fiber activity may down-regulate the efficacy of the particular subpopulation of granule cells or parallel fibers active around that time (when active in the absence of climbing fiber activity, parallel fiber synapses are thought to either remain unchanged or increase in efficacy). The repeated pairing of a particular tone with an air puff, at a particular delay after tone onset, therefore changes the system response in the vicinity of that particular latency. Although some theorists continue to question the relative importance of the cerebellum for eye-blink conditioning (Bloedel 1992; Llinás and Welsh 1993; Gruart et al. 1994; Bloedel and Bracha 1995), and although alternative theories of olivocerebellar function continue to emerge (e.g., Bower and Kassel 1990; Bloedel et al. 1993; Welsh et al. 1995), it is widely agreed that the cerebellar cortex is well-suited to serve as the substrate of motor timing demonstrated in the conditioned eye-blink. Specifically to the case of eye-blink conditioning, the timing-specific reduction of Purkinje cell output is thought to release the interpositus nucleus from inhibition at the appropriate time for response production.

Given that so many of these theories emphasize the involvement of cerebellar cortex, it would be reasonable to expect that the evidence supporting the cerebellar cortex as the site of learning-related changes in eye-blink conditioning is unimpeachable. Surprisingly, this is not the case. In fact, it has proven difficult to demonstrate conclusively a necessary relationship between the cerebellar cortex and the induction of motor learning in the eye-blink conditioning paradigm. In eye-blink conditioning, evidence has instead consistently implicated the deep cerebellar nuclei—the anterior interpositus (aINP), in particular—as integral for the occurrence and maintenance of learning (Thompson 1986; Steinmetz et al. 1992a). Lesions of the aINP eliminate previously conditioned responses (CRs) for as long as 8-10 months (Steinmetz et al. 1992b) and make naive animals untrainable (Lavond et al. 1985; Steinmetz et al. 1992a; Thompson and Steinmetz 1992; Krupa et al. 1993). Lesions of the cerebellar cortex, meanwhile, have had highly variable results on conditioned responses, ranging from little effect (Woodruff-Pak et al. 1993) to destabilization of response timing and amplitude (Lavond and Steinmetz 1989; Perrett et al. 1995) to abolition (Hardiman and Yeo 1992). Given that it is difficult to lesion the cerebellar cortex extensively without causing ancillary damage to the deep nuclei, these lesion data do not strongly support theories that explain conditioning in terms of cortical processing.

Particularly interesting in relation to this issue are electrophysiological studies that implicate the aINP in the maintenance of learning-related neural activity elsewhere in the brain. Modulations of single- and multiple-unit firing rates precede and "model" the closing of the eyelid in the aINP, cerebellar cortex, red nucleus, trigeminal nucleus, motor abducens, thalamic nuclei, basal ganglia, and hippocampus (for review, see Anderson and Steinmetz 1994; Steinmetz 1996). All studies to date, however, suggest that lesions of the aINP eliminate learning-related patterns of firing in other areas. Recent work, for instance, has demonstrated that multiple-unit CR models in the red nucleus and trigeminal complex vanish (along with the CR itself) when the aINP is inactivated by cooling or lidocaine infusion (Chapman et al. 1990; Clark et al. 1992; Clark and Lavond 1996). Similar findings have linked conditioned eye-blink predictive firing in the thalamus and hippocampus to aINP integrity (Sears and Steinmetz 1990; Sears et al. 1996). The aINP models, meanwhile, do not appear to depend on the integrity of the red nucleus (Chapman et al. 1990). These studies do not demnstrate conclusively that plasticity in the aINP is solely responsible for conditioning-related activity, but they certainly demonstrate a repeated failure to prove otherwise.

The literature discussed above does not conclusively rule out the occurrence of non-aINP localized plasticity, however. All studies in which neural activity was seen to be entirely dependent on the aINP involved multiple-unit recording. It is becoming increasingly clear that such recordings may obscure subtle patterns of firing observable in single-unit spike trains. For instance, single-unit recordings in cerebellar cortex made during eye-blink conditioning reveal a subpopulation of Purkinje cells that show CR-related firing-rate decreases ("inhibitory models"); these cells are invisible in multiple-unit records (Gould and Steinmetz 1996). The above studies, then, may have missed conditioning-related activity that survived aINP lesion. It is possible that the relative properties of the population had changed as a result of the lesion, such that inhibitory and excitatory models canceled each other out in the multiple-unit record.

It is also possible that the catastrophic effect that aINP lesions have on eye-blink conditioning (and on extracerebellar firing patterns) may reflect, at least in part, isolation of cerebellar cortical processing. That is, the effect of aINP lesions may not solely reflect intrinsically deep nuclear processing, but may reflect changes in the impact of cortical activity on system output (it has also been suggested that the result of aINP lesions may, in part, reflect changes in the activity of brain stem nuclei; e.g., Bloedel and Bracha 1995). The deep nuclei do serve as a localized "bottleneck" in the outflow of a highly distributed cerebellar cortical system that in itself may be difficult to lesion completely. Cutting of this outflow path could result in massive learning decrements without these decrements entailing destruction of the entire learning substrate. Studies concerning the effects of temporary inactivation on learning and performance of the conditioned eye blink (e.g., Krupa et al. 1993; Nordholm et al. 1993) strongly suggest that at least some of the plastic machinery of learning is intrinsically deep nuclear, but in both studies the researchers noted probable diffusion into the cerebellar cortex, and specifically into the region shown to be involved in eye-blink conditioning (Steinmetz and Sengelaub 1992; Van Ham and Yeo 1992; Gould et al. 1993). It remains possible that the cerebellar cortex (or even some extracerebellar structure) undergoes learning-related plasticity during eye-blink conditioning.

Solid evidence that the cerebellar cortex is intrinsically involved in learning of the conditioned eye-blink requires the documentation of conditioning-related cortical activity recorded in the absence of the driving force that seems to be provided by an intact aINP. Appropriate lesions of the deep nucleus would be expected to eliminate all behavioral traces of conditioning, but it is conceivable that non-naive patterns of single-unit neural activity in response to CS presentation might survive in the cerebellar cortex. These are the data that are presented in this paper.

In this study, rabbits were trained to criterion (>75% CRs) in forward-paired delay conditioning of the eye-blink reflex. After reaching criterion for two sessions, animals received either kainic acid or sham saline lesions of the aINP, and then additional sessions were run in concert with single-unit recording from the overlying cerebellar cortex. Permanent excitotoxic lesions were used because: (1) They spare fibers of passage, such that axon collaterals from the pons and inferior olive are left intact; (2) they have proven effective in eliminating behavioral evidence of conditioning (Lavond et al. 1985); (3) simultaneous inactivation of the aINP and recording from the cerebellar cortex is technically difficult given the anatomical arrangement of the cortical/aINP complex; and (4) it is possible that aINP-controlled conditioning-related firing patterns in cerebellar cortex could briefly survive aINP inactivation—therefore the removal of aINP cells entirely, followed by a 7-day recovery period, tests more strongly the hypothesis that the cerebellar cortex maintains conditioning-related neural activity independently of the aINP.

Our data suggest that the cerebellar cortex does in fact support conditioning-related patterns of neural activity following aINP lesion, but that such activity tends to be less well-organized than that seen in a population of cortical cells under normal conditions. The data also suggest that excitatory conditioning-related activity exists in neurons that, in US-alone trials, produce complex or simple spikes in response to the air puff. We interpret these results as evidence of a distributed system of learning, and suggest that the intact system is required for the operation of appropriate timing mechanisms. Specifically, we argue that the aINP is an integral part of a circuit that organizes the activity of the distributed system into a coherent whole, capable of driving a precisely timed eye-blink CR. The excitatory firing patterns produced during the function of this system seem integrally involved in the production of eye blinks, despite the fact that such activity should inhibit the aINP. The implications of these findings for cerebellar learning theories are discussed below.

ORDER FORM

LEARNING & MEMORY

ISSN 1054-9803

VOL. 4, 1997 (contains 6 bimonthly issues)
- ☐ Individual price (U.S.) - $95
- ☐ Individual price (R.O.W.* Surface Delivery) - $110
- ☐ Individual price (R.O.W. Airlift Delivery) - $130
- ☐ Institutional price (U.S.) - $210
- ☐ Institutional price (R.O.W. Surface Delivery) - $225
- ☐ Institutional price (R.O.W. Airlift Delivery) - $245
- ☐ Please send me a sample issue

*R.O.W. = Rest of world

Personal orders must be prepaid by personal check, credit card or money order.
☐ Check or money order enclosed (U.S. Bank Checks Only), made out to CSHL Press
Charge to: ☐ MASTERCARD ☐ VISA ☐ AMERICAN EXPRESS ☐ DISCOVER

ACCOUNT NO. _____ EXP. _____
SIGNATURE _____ TEL. _____
NAME _____ FAX _____
ADDRESS _____
CITY/STATE/ZIP/COUNTRY _____

Please check title(s) that most closely describe(s) your position:
(Check all that apply)
- ___(1) Professor
- ___(2) Graduate student
- ___(3) Postdoctoral scientist
- ___(4) Lab director
- ___(5) Lab technician
- ___(6) Medical student
- ___(7) Undergraduate student
- ___(8) Librarian
- ___(9) Publisher
- ___(10) Research associate
- ___(17) Principal investigator
- ___(19) M.D.
- ___(29) Purchasing agent/buyer
- ___(99) Other _____

Please check your employment category:
(Please check one)
- ___(1) University/college
- ___(2) Research institute/foundation
- ___(3) Hospital
- ___(4) Medical school
- ___(5) Industry
- ___(6) Government
- ___(7) Library/information center

Please check your primary field of interest:
(Please check one)
- ___(1) Biochemistry
- ___(2) Cell biology
- ___(3) Developmental biology
- ___(4) Epidemiology
- ___(5) Genetics
- ___(6) Immunology
- ___(7) Microbiology
- ___(8) Molecular biology
- ___(9) Neurobiology
- ___(10) Plant biology
- ___(11) Pharmacology
- ___(12) Virology
- ___(13) Oncology
- ___(33) Protein chemistry
- ___(34) Veterinary medicine
- ___(99) Other _____

Please indicate your areas of interest and techniques you use in your research
(Check all that apply)

MOLECULAR BIOLOGY
- ___(1) Autoradiography
- ___(2) cDNA synthesis
- ___(3) Cloning
- ___(4) DNA footprinting
- ___(5) DNA purification
- ___(6) Electroporation
- ___(7) Eukaryotic expression
- ___(8) Genomic mapping and screening
- ___(9) In situ hybridization
- ___(10) In vitro transcription
- ___(11) In vitro translation
- ___(12) Library screening
- ___(13) Mutagenesis
- ___(14) Nonisotopic labeling and detection
- ___(15) Nucleic acid electrophoresis
- ___(16) Nucleic acid radiolabeling
- ___(17) Nucleic acid sequencing
- ___(18) Nucleic acid synthesis
- ___(19) PCR
- ___(20) Prokaryotic expression
- ___(22) Pulsed field electrophoresis
- ___(23) RNA analysis
- ___(24) RNA purification
- ___(25) Software-aided sequence analysis
- ___(26) Transfections

PROTEIN CHEMISTRY
- ___(30) Amino acid analysis
- ___(31) Chromatography
- ___(32) Affinity
- ___(33) Gas
- ___(34) HPLC
- ___(35) Thin layer
- ___(36) Electrophoresis of proteins
- ___(37) Glycoprotein analysis
- ___(38) Peptide mapping
- ___(39) Peptide synthesis
- ___(40) Protein kinase assays
- ___(42) Protein purification
- ___(43) Protein sequencing

IMMUNOLOGY
- ___(50) Antibody labeling
- ___(51) Antibody purification
- ___(52) Ascites production
- ___(53) ELISA
- ___(54) Flow cytometry
- ___(55) Hybridoma production
- ___(56) Immunohistochemistry
- ___(57) Immunoprecipitation
- ___(58) RIA
- ___(59) Western blotting

CELL BIOLOGY
- ___(70) Apoptosis
- ___(71) Cell proliferation
- ___(72) ES cell culture
- ___(73) Growth factors/cytokines
- ___(74) Insect cell culture
- ___(75) Mammalian cell culture
- ___(76) Microscopy
- ___(77) Plant cell culture
- ___(78) Receptor studies
- ___(79) Serum-free cell culture

TEL: Continental U.S. and Canada: 1-800-843-4388   All other locations: 516-349-1930   FAX: 516-349-1946

Orders for Japan: Maruzen Company Ltd., 3 - 10, Nihonbashi 2-Chome, Chuo-ku, Tokyo 103, Japan. **Orders for India, Pakistan, Bangladesh, Sri Lanka, and Nepal:** Narosa Book Distributors Private Limited, 6 Community Centre, Panchsheel Park, New Delhi 110 017, India.

For fastest service, call:
1-800-843-4388
Continental U.S. and Canada
516-349-1930
All other locations
FAX: 516-349-1946
E-MAIL: cshpress@cshl.org
world wide web site
http://www.cshl.org/

Cold Spring Harbor Laboratory Press
Plainview, NY 11803-2500

NO POSTAGE
NECESSARY
IF MAILED
IN THE
UNITED STATES

BUSINESS REPLY MAIL
FIRST CLASS MAIL   PERMIT NO. 150   HICKSVILLE, NY

POSTAGE WILL BE PAID BY ADDRESSEE

Cold Spring Harbor Laboratory Press
10 Skyline Drive
Plainview, NY 11803-2500

NO POSTAGE
NECESSARY
IF MAILED
IN THE
UNITED STATES

BUSINESS REPLY MAIL
FIRST CLASS MAIL   PERMIT NO. 150   HICKSVILLE, NY

POSTAGE WILL BE PAID BY ADDRESSEE

Cold Spring Harbor Laboratory Press
10 Skyline Drive
Plainview, NY 11803-2500

For fastest service, call:
1-800-843-4388
Continental U.S. and Canada
516-349-1930
All other locations
FAX: 516-349-1946
E-MAIL:
cshpress@cshl.org
world wide web site
http://www.cshl.org

**Cold Spring Harbor
Laboratory Press**
Plainview, NY 11803-2500

Materials and Methods

SUBJECTS

Nineteen male, New Zealand white rabbits (Myrtle's Rabbitry, Thompson Station, TN) weighing >1.75 kg at the start of training served as the subjects in this experiment. Twelve eventually served as kainic acid-lesioned subjects and the remaining seven became sham-lesioned controls. Before and between sessions, animals were housed individually in clean cages with ad lib access to water and food. The housing area maintained rabbits on a 12hr light–dark schedule.

PRESURGERY TRAINING

Naive rabbits were adapted to the sound-attenuating conditioning chamber, and the Plexiglas restraint box placed inside the chamber, for 90 min (spread over two sessions). The conditioning chamber was equipped with a fan that produced background noise (~59 dB) and a speaker to deliver the tone CS. In the restraint box, a metal plate that fit just over the rabbit's head was equipped to hold the air puff US/eye-blink monitor assembly. This assembly held a nozzle attached to an air tank that delivered the air puff US, and also an infrared emitter/detector that measured changes in diffraction of a beam of infrared light (Thompson et al. 1994); these measured changes were amplified and calibrated to reflect millimeters of eye closure in response to unconditioned and conditioned stimuli. The entire assembly hung directly (4-5 mm) in front of the animal's left eye during conditioning.

On the third day, forward-paired delay conditioning sessions began, one per day. Sessions included 12 blocks of 10 trials each: nine pairings of a 350 msec CS tone [1 kHz, 85 dB sound pressure level (SPL)] coterminating with a 100-msec US (3 psi) air puff directed at the eye; one CS-alone trial in which no air puff was delivered. The air puff strength was sufficient to produce a reliable reflex eye blink in all animals. An additional block of 10 US-alone trials was presented at the beginning and end of each session. Trials were separated by an intertrial interval (ITI) that varied randomly between 20 and 30 sec.

Animals were trained in this fashion until they produced CRs that is, >0.5-mm blinks that began before air puff delivery, on 75% of the paired trials in two sessions.

SURGERY

Surgeries were performed under aseptic conditions. Rabbits were anesthetized with a combination of ketamine [60 mg/kg i.m. (intramuscularly)] and xylazine [6 mg/kg s.c. (subcutaneouly)], and maintained throughout surgery on 1 ml i.m. injections of a ketamine/xylazine mixture administered every 45 min. Once anesthetized, rabbits were secured in a standard stereotaxic head holder, and the skull over the left cerebellum was drilled away.

Lesions were placed in the aINP via microinjections of kainic acid (0.2 µl of 25 nmoles of kainic acid per microliter of saline), according to standard coordinates reported in Lavond et al. (1985). A Hamilton syringe containing the infusate was lowered through the skull hole (+0.7 AP and +5.5 ML relative to λ) to a depth of 14.5 mm below λ, where it sat untouched for 2 min. The 0.2 µl of solution was then pressure injected over the course of 8 min. After an additional 2 min during which the acid was allowed to diffuse away from the needle tip, the syringe was raised out of the skull over a 1-min period. A sham-lesioned control group received similar penetrations, but normal saline was infused instead of kainic acid.

After lesioning, the skull hole was filled with bone wax, and a cylindrical base capable of supporting a micromanipulator was cemented over Larsell's Lobule H-VI of the cerebellar cortex (H-VI) with dental acrylic; the base was filled with petroleum jelly to protect the brain surface. Two head bolts were cemented anterior to the base to facilitate securing the rabbit's head during recording sessions, and also as an anchor from which to hang the air puff/eyelid measurement assembly. The scalp was then sutured around the headstage, and povidone–iodine ointment was applied liberally to the wound. In-cage water was replaced with sulmet for the 5 days following surgery. Rabbits were allowed one full week of recovery before training resumed.

RECORDING

During the five sessions of post-surgery training, single-unit activity was recorded from the cerebellar cortex. An acid-etched, varnish-insulated stainless-steel rod (3–5 MΩ) was set in a hydraulic micromanipulator that was then attached to the base cemented into the headstage. As the electrode was lowered into the cortex, cells were isolated auditorially and visually (using a 2-to-1 signal-to-noise criterion). Amplitude-discriminated action potentials were sent as TTL pulses to the run-time computer, and stored to computer disk, whereas the waveform was saved onto VCR tape; more recently, a commercially available hardware/software package (Spike2, from CED, Ltd; Cambridge, England) allowed us to digitize raw neural signals, and to store them directly onto a computer hard disk. After collection of data during two to four blocks of trials, the electrode was lowered until a new cell was isolated, and the process was repeated. A subset of neurons were held through an additional one or two blocks of US-alone trials. The last penetration was marked with a current lesion (100 microamps for 10 sec), as was any penetration in which more than two obviously conditioning-related neurons were isolated, and the locations of lesions were noted.

HISTOLOGY

Following the final session, each rabbit was overdosed with an intravenous injection of pentobarbitol (4 ml) and perfused transcardially with saline and formalin. The brains were removed, fixed in a sucrose formalin solution, embedded, sliced coronally to 40-µg thickness, stained with Prussian blue to obtain a precise measure of electrode placement, and counterstained with cresyl violet for visualization of cell bodies.

Lesion effectiveness was measured both behaviorally and through examination of slices in the planes 0.5, 1.0, and 1.5 mm anterior to λ. The aINP was visualized under a light microscope at 400× magnification, and the number of neurons in the dorsal aspect of the nucleus was counted. A similar count was made of the contralateral aINP for direct within-rabbit comparison. Centering of the microscope field was done in relation to anatomical landmarks in a standardized fashion across hemispheres and animals. Neuron counts were based on standard morphological criteria, including cell size, shape, and nonhomogeneity of staining.

DATA ANALYSIS

Peristimuli time histograms of neural activity, constructed from amplitude-discriminated spike counts or from discriminated spike trains produced via template matching algorithms on the Spike2 system, were organized into time bins. Each of three same-length trial periods—the time before CS onset (pre-CS period), the time between CS and US onsets (CS period), and the time after US onset (US period)—were divided into five time bins. Spike count standardization procedures established whether significant learning-related activity occurred. Briefly, firing rates within bins in the CS and US periods were compared with appropriate bins in the pre-CS period, and paired t-tests on the different scores provided the standardized firing rate. Significant trial-related activity was recognizable as t-scores less likely to occur by chance than 5%; corrections that took the pre-CS period variability into account limited the likelihood of type-1 errors.

Cells showing significant activity levels within a trial were categorized using the following scheme: (1) CS-onset cells, showing phasic bursts of activity briefly following and sharply time-locked to tone presentation; (2) US-onset cells, showing phasic bursts of activity briefly following air puff presentation; (3) movement-related cells, showing increased activity after the onset of eyelid closure; and (4) conditioning-related cells, showing increases or decreases in firing rates between CS and US or blink onset that were not time-locked to either of the stimuli, and that showed broader firing rate peaks. Typically, eye-blink generative responses show significant changes in firing rate between 60 and 30 msec before blink onset and model the CR; in this case, however, there was expected to be no post-lesion CRs for the activity to model. Therefore, this slightly more indirect method was used for classification.

Descriptive analyses were used to summarize the number of cells showing learning-related activity in the sessions immediately following lesion of the aINP. Subsequent χ^2 tests (along with related effect-size measurements) and t-tests revealed (1) whether learning-related activity found in the cerebellar cortex in the absence of the aINP changed with time post-lesion; (2) whether the amount of activity found in sham-lesioned animals was similar to that found in aINP-lesioned animals; and (3) whether the distributions of single-cell response timings, and the proportion of firing-rate increases and decreases differed between the lesioned and nonlesioned groups. For the purpose of these analyses, the peristimulus time histograms were transformed into firing rate running averages. Standardized firing rates (standardized this time on the entire pre-CS period variability) were smoothed on a Gaussian curve and the result—a relatively smoothly changing curve—was plotted as a function of time. Onsets and peaks of activity were easily identifiable from these plots.

Results

BEHAVIORAL EFFECTS OF KAINIC ACID aINP LESIONS

Figure 1 shows how kainic acid infusion into the aINP affected conditioning. On the *x*-axis are the last 4 days of presurgery training and the 5 days of postsurgery training for both kainic acid- and sham-lesioned groups. As has been observed repeatedly in earlier studies (Lavond et al. 1985; Thompson 1988; Clark et al. 1992; Steinmetz et al. 1992a,b; Krupa et al. 1993; Bracha et al. 1994), aINP lesions effectively reduced the occurrence of eye-blink CRs to chance blinking levels.

A 2 × 5 mixed-effect ANOVA with group as the two-level between-group effect and post-surgery session as the five-level within-group effect confirmed that the kainic acid- and sham-lesioned groups performed to significantly different levels of conditioning following surgery, [$F(1,16) = 45.3$; $P < 0.001$]. A similar 2 × 4 ANOVA on pre-surgery sessions showed only an effect of session [$F(3,36) = 30.7$; $P < 0.001$] and not of group ($F < 1$). After surgery the sham-lesioned animals did show a temporary decrease in level of conditioning (interaction of group with session, [$F(4,64) = 4.664$; $P < 0.005$] that was probably attributable either to tiny lesions caused by the pressure injections themselves or to the week after surgery spent untouched in the home cages; by post-surgery day 5, however, the sham group performed indistinguishably from the last day before surgery ($t < 1$).

The proportion of CRs made per session in the kainic acid-lesioned group (collapsed across days) was also compared by *t*-test with that predictable due to spontaneous blink rates (five per session). The difference (which was virtually nonexistent) did not approach significance ($t < 1$). The lesions eliminated the animals' ability to perform conditioned eye blinks. This pattern was consistent for all but one rabbit in the kainic acid-lesioned group.

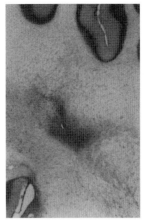 

Figure 2: Anatomical results of excitotoxic lesions. Photomicrographs of a single kainic acid-lesioned animal's lesioned (*left*) and control (*right*) aINPs.

ANATOMICAL EXAMINATION OF LESION EFFECTIVENESS

Figure 2 presents a pair of photomicrographs, showing the left and right interpositus–dentate regions from a kainic acid-lesioned rabbit. Notice the near absolute neuron loss in the left aINP and dentate nuclei, and its replacement with heavy accumulations of glial cells.

Neurons were counted in single microscope fields centered on the medial aspect of the dentate-interpositus triangle and set to magnify the tissue 400×. Counts were taken from injected (left) hemispheres and from the same location in control hemispheres of representative sections (~0.5, 1.0, and 1.5 mm anterior to λ); the average (and standard error) cell counts for both groups are dis-

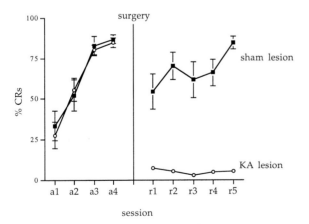

Figure 1: Percent CRs made by kainic acid-lesioned and sham-lesioned animals in the last four sessions of presurgery training (a1–a4) and in the five sessions of postsurgery training (r1–r5). Bars, standard error.

played in Figure 3. Comparisons of experimental and control hemispheres confirmed what is obvious from the behavioral results and from Figure 2, that kainic acid decimated the region of deep cerebellar nucleus traditionally considered to be essential for eye-blink conditioning.

The difference between groups was analyzed by computing the proportion of cells spared by the lesion. That is, the number of neurons counted within the injected (left) aINP was divided by the number of neurons counted within intact (right) aINP. These proportions were than compared between groups. Proportions were calculated, rather than simple differences, to control for the possibility that cells would be lost: (1) in the sham-lesioned hemispheres as a result of the saline injection; and (2) in control hemispheres as a result of connections between the left and right cerebella. If there was some cell loss in both groups, and in both hemispheres of the kainic acid-lesioned animals, then simple differences between left and right hemisphere cell counts might cause an underestimation of the effect of kainic acid infusion. The reported results were replicated in difference measurements and absolute cell number measurements, however; in fact, the small difference between cell counts in the control hemispheres of the kainic acid- and sham-lesioned groups proved insignificant ($t < 1$).

In sham-lesioned animals, the left-to-right cell count proportion averaged 74%, whereas in kainic acid-lesioned animals, it averaged 17%. The difference between groups on this measure was significant [$t(16) = -4.33$; $P < 0.001$]. The impact of kainic acid lesions was to more than quadruple the cell loss attributable to the injection.

It is clear that even the saline injection destroyed some cells; the left-to-right cell count proportion in control animals was significantly smaller than unity [$t(6) = -3.72$; $P < 0.001$].

CORRELATION OF CELL LOSS AND CR EXECUTION IN LESIONED ANIMALS

The variability in kainic acid-related cell sparing proportions was related to the proportion of post-surgery trials in which eye-blink CRs were observed. As expected, the two measures correlated positively ($r = 0.78$). This correlation was largely driven by one animal that was an outlier on both measures, showing fewer cells destroyed and higher proportion of CRs than the mean. With this animal removed from the analysis, the correlation between cell loss and CR likelihood became insignificant, and in fact slightly negative. The lack of correlation reflects the lack of variability (i.e., the floor effect) seen in percent CRs produced (and to some extent in the number of cells found). The few cells remaining were incapable of driving any sort of CR, and thus there was no correlation between behavioral performance and cell proportion. The animal with spared cells was excluded from subsequent analyses comparing the sham- and kainic acid-lesioned group. Exclusion of this animal from the cell count analysis reduced the average left-to-right cell sparing proportion to 10%.

SINGLE-UNIT RESPONSES IN KAINIC ACID- AND SHAM-LESIONED ANIMALS

In total, recordings were taken from 293 neurons in cerebellar cortex. Examination of electrode tracks and lesions revealed that the majority of these neurons were located in Larsell's lobule H-VI or the anterior lobe directly anterior to H-VI. Figure 4 shows the general region from which all neurons were recorded, and Figure 5 shows a representative H-VI slice, with an electrode tract and marking lesion indicated.

In analyses published previously of cerebellar single-unit activity related to eye-blink conditioning (e.g., Berthier and Moore 1986; Gould and Steinmetz 1996), neurons were classified as either nonmodulated in the conditioning context, responsive to the onsets of stimuli, or conditioning-

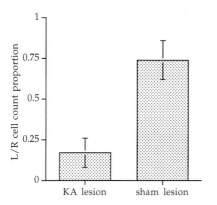

Figure 3: Average (and standard errors) of cell sparing, measured as the left to right proportion of cell numbers counted in the aINP, for the kainic acid-lesioned and sham-lesioned groups.

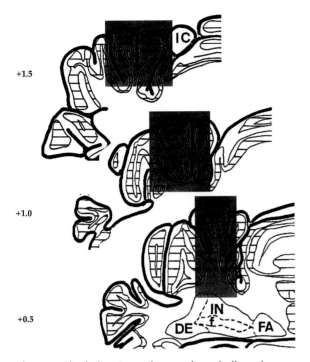

Figure 4: Shaded regions of coronal cerebellar schematics showing the region in which most electrode tracks were located; numbers to the *left* represent distance anterior of λ. (DE) Dentate nucleus; (f) fiber tract; (FA) fastigial nucleus; (IC) inferior colliculus; (IN) interpositus nucleus.

related in sham- and kainic acid-lesioned samples. Effective lesions of the aINP did not eliminate the occurrence of conditioning-related firing patterns in single neurons.

CORRELATION BETWEEN aINP CELL LOSS AND CONDITIONING-RELATED SINGLE-UNIT ACTIVITY

It is important to assess the possible contribution of spared aINP to the activity observed in kainic acid-lesioned animals. Therefore, the raw number of cells remaining in the lesioned aINP of each kainic acid-lesioned animal was related to the percentage of recorded neurons that had been classified as conditioning-related in that same animal. A striking lack of relationship between these measurements was present ($r = 0.04$). The subject with the least spared INP provided the highest proportion of conditioning-related neurons, and the next highest proportion of conditioning-related neurons came from an animal with a relatively spared aINP. This latter subject is the animal, discussed above, that produced a sizeable number of CRs; with this

related. We began our neural analysis by preparing the same descriptions, according to the criteria set forth in the Materials and Methods. Figure 6A presents the classificatory breakdown for the 110 neurons recorded from the cerebellar cortex of sham-lesioned animals. The proportion of neurons in each category (nonmodulated, CS-onset, US-onset, onset of both stimuli, and conditioning-related) largely replicates the results of the above-mentioned studies. Slightly under 25% of neurons recorded in the vicinity of H-VI showed patterns of activity that were related to the behavioral CR. This replication serves to validate the slight differences in statistical methods used to quantify the firing patterns.

Figure 6B presents a similar breakdown for the 183 neurons recorded from the cerebellar cortices of kainic acid-lesioned animals. The similarity between Figure 6 A and B, is such that the graphs may look identical under casual examination. We found no statistical differences between the groups in terms of the proportion of neurons that fit each general classification. Most importantly, similar proportions of cells appeared to be conditioning-

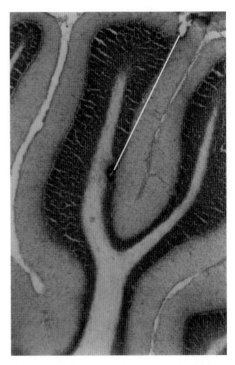

Figure 5: Coronal section through left Larsell's lobule H-VI showing representative electrode track. The white line from entry mark to lesion mark is the path in which the electrode traveled; note that the track skims the Purkinje cell layer.

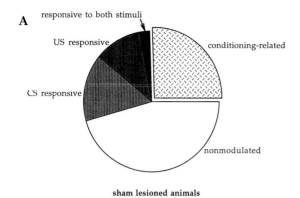

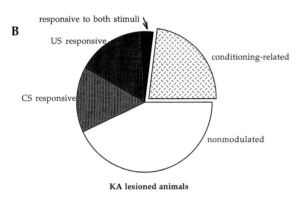

Figure 6: Breakdown of neurons recorded into those not modulated by conditioning trials, those responsive to the CS, the US, or both stimuli, and those designated as conditioning-related. (A) Sham-lesioned animals; (B) kainic acid-lesioned animals.

subject's data removed, the correlation between spared aINP and the probability of recorded neurons being conditioning-related became somewhat (but not significantly) negative, $r = -0.52$. Although this result is surprising and difficult to interpret, it certainly re-affirms the point that the conditioning-related activity patterns observed were not supported by spared aINP.

POPULATION DISTRIBUTION PROPERTIES OF CONDITIONING-RELATED ACTIVITY IN KAINIC ACID-LESIONED AND SHAM-LESIONED ANIMALS

Closer examination of the subsamples of neurons classified as conditioning-related revealed substantial differences between the experimental and control groups. Kainic acid lesions changed the proportions of neurons showing firing rate inhibition relative to excitation. Researchers have reported previously that conditioning-related firing patterns seem to fall into two types, those that show increases in firing rate preceding CR-onset (for the remainder of the paper, these will be referred to as "on" cells), and those that show decreases in firing rate preceding CR-onset ("off" cells). In agreement with Berthier and Moore (1986), we have typically found that excitatory cells outnumber inhibitory cells by a ratio of ~2:1 (e.g., Gould and Steinmetz 1996). As shown in the left half of Figure 7, this result was replicated in our sham-lesioned group. Figure 8 shows examples of summary peri-stimulus time histograms of these two basic types of neural responses. The right half of Figure 7, however, reveals that on cells and off cells were equally likely after kainic acid lesion. The likelihood of an isolated neuron being an on cell was significantly different for the two groups [$t(16) = 2.46, P < 0.03$].

In addition, a type of firing profile that could not be classified as simply increasing or decreasing was observed in neurons from kainic acid-lesioned animals, and not from those in sham-lesioned animals. These neurons were reliably bi- or triphasic, turning, for example, off and then on. Figure 9 shows a traditional on cell found in a kainic acid-lesioned animal, as well as a more unusual triphasic discharging neuron. The kainic acid lesion changed the shapes of individual firing patterns observed, as well as the proportion of on cells to off cells.

These differences were studied in finer detail. The modulations of firing rate were characterized roughly with regard to the onsets and peaks of

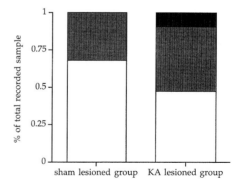

Figure 7: Kainic acid lesions of the aINP changed properties of observed single-unit cortical firing patterns. Proportions of the conditioning-related subpopulation of recorded neurons that showed excitatory modulation (white bars), inhibitory modulation (lightly shaded bars), or some combination of the two, that is, bi- and triphasic cells, (darkly shaded bars) for kainic acid- and sham-lesioned groups.

Anderson, B.J. and J.E. Steinmetz. 1994. Cerebellar and brainstem circuits involved in classical eyeblink conditioning. *Rev. Neurosci.* **5:** 1–23.

Berthier, N.E. and J.W. Moore. 1986. Cerebellar Purkinje cell activity related to the classically conditioned nictitating membrane response. *Exp. Brain Res.* **63:** 341–350.

Bloedel, J.R. 1992. Functional heterogeneity with structural homogeneity: How does the cerebellum operate? *Behav. Brain Sci.* **15:** 666–678.

Bloedel, J.R. and V. Bracha. 1995. On the cerebellum, cutaneomuscular reflexes, movement control and the elusive engrams of memory. *Behav. Brain res.* **68:** 1–44.

Bloedel, J.R., V. Bracha, and P.S. Larson. 1993. Real time operations of the cerebellar cortex. *Can. J. Neurol. Sci.* **20 (Suppl. 3):** S7–S18.

Bower, J.M. and J. Kassel. 1990. Variability in tactile projection patterns to cerebella folia crus IIA of the Norway rat. *J. Compar. Neurol.* **302:** 768–778.

Bracha, V. and J.R. Bloedel. 1996. The multiple-pathway model of circuits subserving the classical conditioning of withdrawal reflexes. In *Acquisition of motor behavior in vertebrates* (ed. J.R. Bloedel, T. Ebner, and S. Wise), pp. 175–204. MIT Press, Cambridge, MA.

Bracha, V., M.L. Webster, N.K. Winters, K.B. Irwin, and J.R. Bloedel. 1994. Effects of muscimol inactivation of the cerebellar interposed-dentate nuclear complex on the performance of the nictitating membrane response in the rabbit. *Exp. Brain Res.* **100:** 453–468.

Buisseret-Delmas, C. and P. Angaut. 1989. Anatomical mapping of the cerebellar nucleocortical projections in the rat: A retrograde labeling study. *J. Compar. Neurol.* **288:** 297–310.

Buonomano, D.V. and M.D. Mauk. 1994. Neural network model of the cerebellum: Temporal discrimination and the timing of motor responses. *Neural Computat.* **6:** 38–55.

Chapeau-Blondeau, F. and G. Chauvet. 1991. A neural network model of the cerebellar cortex performing dynamic associations. *Biol. Cybernetics* **65:** 267–279.

Chapin, J.K. and M.A.L. Nicolelis. 1995. Beyond single unit recording: Characterizing neural information in networks of simultaneously recorded neurons. In *Scale in conscious experence: Is the brain too important to be left to specialists to study?* (ed. J. King and K.H. Pribram), pp. 135–153. Erlbaum Associates, Hillsdale, NJ.

Chapman, P.F., J.E. Steinmetz, L.L. Sears, and R.F. Thompson. 1990. Effects of lidocaine injection in the interpositus nucleus and red nucleus on conditioned behavioral and neuronal responses. *Brain Res.* **537:** 149–156.

Chrobak, J.J. and G. Buzsáki. 1996. High-frequency oscillations in the output networks of the hippocampal-entorhinal axis of the freely behaving rat. *J. Neurosci.* **16:** 3056–3066.

Clark, R.E. and D.G. Lavond. 1996. Neural unit activity in the trigeminal complex with interpositus or red nucleus inactivation during classical eyeblink conditioning. *Behav. Neurosci.* **110:** 13–21.

Clark, R.E., A.A. Zhang, and D.G. Lavond. 1992. Reversible lesions of the cerebellar interpositus nucleus during acquisition and retention of a classically conditioned behavior. *Behav. Neurosci.* **106:** 879–888.

Cohen, J. 1992. A power primer. *Psycholog. Bull.* **112:** 155–159.

Dow, R.S. and G. Moruzzi. 1958. *The physiology and pathology of the cerebellum* University of Minnesota Press, Minneapolis, MN.

Gilman, S., J.R. Bloedel, and R. Lechtenberg. 1981. *Disorders of the cerebellum* F.A. Davis, Philadelphia, PA.

Gould, T.J. and J.E. Steinmetz. 1996. Changes in rabbit cerebellar cortical and interpositus nucleus activity during acquisition, extinction and backward classical eyelid conditioning. *Nerobiol. Learn. Mem.* **65:** 17–34.

Gould, T.J., L.L. Sears, and J. Steinmetz. 1993. Possible CS and US pathways for rabbit classical eyelid conditioning: Electrophysiological evidence for projections from the pontine nuclei and inferior olive to cerebellar cortex and nuclei. *Behav. Neural Biol.* **60:** 172–185.

Grossberg, S. and J.W.L. Merrill. 1992. A neural network model of adaptively timed reinforcement learning and hippocampal dynamics. *Cognit. Brain Res.* **1:** 3–38.

Gruart, A., P. Blazquez, and J.M. Delgado-Garcia. 1994. Kinematic analyses of classically-conditioned eyelid movements in the cat suggest a brain stem site for motor learning. *Neurosci. Lett.* **175:** 81–84.

Hardiman, M.J. and C.H. Yeo. 1992. The effect of kainic acid lesions of the cerebellar cortex on the conditioned nictitating membrane response in the rabbit. *Eur. J. Neurosci.* **4:** 966–980.

Ito, M. 1993. Movement and thought: Identical control mechanisms by the cerebellum. Reply to Leiner, Leiner and Dow. *Trends Neurosci.* **16:** 448–450.

Kawato, M. and H. Gomi. 1992. The cerebellum and VOR/OKR learning models. *Trends Neurosci.* **15:** 445–453.

Kirschfeld, K., R. Feiler, and F. Wolf-Oberhollenzer. 1996. Cortical oscillations and the origin of express saccades. *Proc. R. Soc. Lond. B. Sci.* **263:** 459–468.

Krupa, D.J., J.K. Thompson, and R.F. Thompson. 1993. Localization of a memory trace in the mammalian brain. *Science* **260:** 989–991.

Lavond, D.G. and J.E. Steinmetz. 1989. Acquisition of classical conditioning without cerebellar cortex. *Behav. Brain Res.* **33:** 113–164.

Lavond, D.G., T.L. Hembree, and R.F. Thompson. 1985.

Effect of kainic acid lesions of the cerebellar interpositus nucleus on eyelid conditioning in the rabbit. *Brain Res.* **326:** 179–182.

Lavond, D.G., C.G. Logan, J.H. Sohn, W.D.A. Garner, and S.A. Kanzawa. 1990. Lesions of the cerebellar interpositus nucleus abolish both nictitating membrane and eyelid EMG conditioned responses. *Brain Res.* **514:** 238–248.

Lisberger, S.G. and T.J. Sejnowski. 1992. Motor learning in a recurrent network model based on the vestibulo-ocular reflex. *Nature* **360:** 159–161.

Llinas, R. and J.P. Welsh. 1993. On the cerebellum and motor learning. *Curr. Opin. Neurobio.* **3:** 958–965.

Marr, D. 1969. A theory of cerebellar cortex. *J. Physiol.* **202:** 437–470.

Moore, J.W. and D.E.J. Blazis. 1989. Simulation of a classically conditioned response: A cerebellar neural network implementation of the Sutton-Barto-Desmond model. In *Neural models of plasticity: Experimental and theoretical approaches* (ed. J.H. Byrne and W.O. Berry), pp. 18–207. Academic Press, San Diego, CA.

Moyer, R.R., Jr., R.A. Deyo, and J.F. Disterhoft. 1990. Hippocampectomy disrupts trace eye-blink conditioning in rabbits. *Behav. Neurosci.* **104:** 243–252.

Nordholm, A.F., J.K. Thompson, C. Dersarkissian, and R.F. Thompson. 1993. Lidocaine infusion in a critical region of cerebellum completely prevents learning of the conditioned eyeblink response. *Behav. Neurosci.* **107:** 882–886.

Perrett, S.P., B.P. Ruiz, and M.D. Mauk. 1993. Cerebellar cortex lesions disrupt learning-dependent timing of conditioned eyelid responses. *J. Neurosci.* **13:** 1708–1718.

Rosenthal, R. and R.L. Rosnow. 1991. *Essentials of behavioral research: Methods and data analysis* (2nd ed.) McGraw-Hill, New York, NY.

Schmajuk, N.A. and J.W. Moore. 1985. Real-time attentional models for classical conditioning and the hippocampus. *Physiolog. Psychol.* **13:** 278–290.

Schreurs, B.G., M.M. Oh, C. Hirashima, and D.L. Alkon. 1995. Conditioning-specific modification of the rabbit's unconditioned nictitating membrane response. *Behav. Neurosci.* **109:** 24–33.

Sears, L.L., S.F. Logue, and J.E. Steinmetz. 1996. Involvement of the ventrolateral thalamic nucleus in rabbit classical eyeblink conditioning. *Behav. Brain Res.* **74:** 105–117.

Sears, L.L. and J.E. Steinmetz. 1990. Acquisition of classically conditioned-related activity in the hippocampus is affected by lesions of the cerebellar interpositus. *Behav. Neurosci.* **104:** 681–692.

———. 1991. Dorsal accessory inferior olive activity diminishes during acquisition of the rabbit classically conditioned eyelid response. *Brain Res.* **545:** 114–122.

Steinmetz, J.E. 1996. The brain substrates of classical eyeblink conditioning in rabbits. In *Acquisition of motor behavior in vertebrates* (ed. J.R. Bloedel, T. Ebner, and S. Wise), pp. 91–114. MIT Press, Cambridge, MA.

Steinmetz, J.E. and D.R. Sengelaub. 1992. Possible conditioned stimulus pathway for classical eyelid conditioning in rabbits. *Behav. Neural Biol.* **57:** 103–115.

Steinmetz, J.E., L.L. Sears, M. Gabriel, Y. Kubota, A. Poremba, and E. Kang. 1991. Cerebellar interpositus nucleus lesions disrupt classical nictitating membrane conditioning but not discriminative avoidance learning in rabbits. *Behav. Brain Res.* **45:** 71–80.

Steinmetz, J.E., D.G. Lavond, D. Ivkovich, C.G. Logan, and R.F. Thompson. 1992a. Disruption of classical eyelid conditioning after cerebellar lesions: Damage to a memory trace system or a simple performance deficit? *J. Neurosci.* **12:** 4403–4426.

Steinmetz, J.E., S.F. Logue, and S.S. Steinmetz. 1992b. Rabbit classically conditioned eyelid responses do not reappear after interpositus nucleus lesion and extensive post-lesion training. *Behav. Brain Res.* **51:** 103–114.

Thompson, R.F. 1986. The neurobiology of learning and memory. *Science* **233:** 941–947.

———. 1988. The neural basis of basic associative learning of discrete behavioral responses. *Trends Neurosci.* **11:** 152–155.

Thompson, R.F. and J.E. Steinmetz. 1992. The essential memory trace circuit for a basic form of associative learning. In *Learning and memory: The behavioral and biological substrates.* (ed. I. Gormezano and E.A. Wasserman), pp. 369–386. Lawrence Erlbaum Associates, Hillsdale, NJ.

Thompson, L.T., J.R. Moyer Jr., E. Akase, and J.F. Disterhoft. 1994. A system for quantitative analysis of associative learning. Part 1. Hardware interfaces with cross-species applications. *J. Neurosci. Methods* **54:** 109–117.

Vaadia, E., I. Haalman, M. Abeles, H. Bergman, Y. Prut, H. Slovin, and A. Aertsen. 1995. Dynamics of neuronal interactions in monkey cortex in relation to behavioural events. *Nature* **373:** 515–518.

Van Ham, J.J. and C.H. Yeo. 1992. Somatosensory trigeminal projections to the inferior olive, cerebellum and other precerebellar nuclei in rabbits. *Eur. J. Neurosci.* **4:** 302–317.

Welsh, J.P., E.J. Lang, I. Sugihara, and R. Llinas. 1995. Dynamic organization of motor control within the olivocerebellar system. *Nature* **374:** 453–457.

Woodruff-Pak, D.S., D.G. Lavond, C.G. Logan, J.E. Steinmetz, and R.F. Thompson. 1993. Cerebellar cortical lesions and reacquisition in classical conditioning of the nictitating membrane response in rabbits. *Brain Res.* **608:** 67–77.

Received February 18, 1997; accepted in revised form April 25, 1997.

Effect of Varying the Intensity and Train Frequency of Forelimb and Cerebellar Mossy Fiber Conditioned Stimuli on the Latency of Conditioned Eye-Blink Responses in Decerebrate Ferrets

Pär Svensson,[1] Magnus Ivarsson, and Germund Hesslow

Department of Physiology and Neuroscience
Lund University
S-223 62 Lund, Sweden

Abstract

To study the role of the mossy fiber afferents to the cerebellum in classical eye-blink conditioning, in particular the timing of the conditioned responses, we compared the effects of varying a peripheral conditioned stimulus with the effects of corresponding variations of direct stimulation of the mossy fibers. In one set of experiments, decerebrate ferrets were trained in a Pavlovian eye-blink conditioning paradigm with electrical forelimb train stimulation as conditioned stimulus and electrical periorbital stimulation as the unconditioned stimulus. When stable conditioning had been achieved, the effect of increasing the intensity or frequency of the forelimb stimulation was tested. By increasing the intensity from 1 to 2 mA, or the train frequency from 50 to 100 Hz, an immediate decrease was induced in both the onset latency and the latency to peak of the conditioned response. If the conditioned stimulus intensity/frequency was maintained at the higher level, the response latencies gradually returned to preshift values. In a second set of experiments, the forelimb stimulation was replaced by direct train stimulation of the middle cerebellar peduncle as conditioned stimulus. Varying the frequency of the stimulus train between 50 and 100 Hz had effects that were almost identical to those obtained when using a forelimb conditioned stimulus. The functional meaning of the latency effect is discussed. It is also suggested that the results support the view that the conditioned stimulus is transmitted through the mossy fibers and that the mechanism for timing the conditioned response is situated in the cerebellum.

Introduction

An important feature of the classical or Pavlovian learning paradigm is the precise temporal regulation of the conditioned response (CR). In classical eye-blink conditioning, where an animal has learned to blink in response to a conditioned stimulus (CS), such as a tone or electrical skin stimulation, the CR is timed so that the eye is maximally closed and, hence, under natural circumstances, protected at the time of the unconditioned stimulus (US), usually an air puff or a periorbital electrical stimulus. This holds over a wide variety of experimental situations. If the interval between the CS and the US is changed, the animal will learn to change the latency of the CR, so that its protective effect is restored (for references, see Mackintosh 1974). If rabbits are trained with two different tone frequencies as CSs, each tone being followed by the US at a different interval, the animals will learn to emit two CRs with latencies appropriate to each CS (Mauk and Ruiz 1992).

An exception to this rule is the shift in CR onset latency that occurs when the animals have

[1]Corresponding author.

been trained to a particular CS and the intensity or the frequency of this CS is changed. For instance, Pavlov reported that if the frequency of a CS consisting of rhythmic tactile stimulation was increased, a conditioned salivation response appeared earlier (Pavlov 1927, p. 100). More recently, it has been shown that varying the intensity of a tone CS can affect the latency of a conditioned nictitating membrane response (Frey 1969; Leonard and Monteau 1971). A limitation of these studies is that they could not exclude that the latency effects were merely secondary to increases in the CR amplitude. A gradual change of the CS intensity during an individual trial has also been shown to change the CR latency. A gradual increase from, as well as a decrease to, the training intensity caused shorter CR latencies (Kehoe et al. 1995).

The aim of the present study was to determine whether the effect on CR latency of changing the intensity of a peripheral CS could be mimicked when using direct stimulation of the mossy fibers to the cerebellum as the CS. Several lines of converging evidence now suggest that the cerebellum plays a key role in the learning as well as the performance of classically conditioned somatic responses. For instance, a large number of studies have shown that lesions to the anterior interpositus nucleus and of the cerebellar cortex (for review, see Yeo 1991; Thompson and Krupa 1994), as well as pharmacological (Krupa et al. 1993) or physiological (Hesslow 1994) inactivation severely interfere with eye-blink conditioning. Several investigators have suggested that conditioning might utilize mechanisms that have been proposed previously by Brindley (1964), Marr (1969), and Albus (1971) for other forms of motor learning, namely that climbing fibers to the cerebellum can modify the responsiveness of Purkinje cells to mossy fiber/parallel fiber input (for review, see Ito 1984). In the context of classical conditioning this would mean that the CS activates mossy fibers and the US climbing fibers (cf. Fig 1). Several experimental results support this suggestion. For instance, it has been shown that stimulation of the pontine nuclei, the source of a large group of mossy fibers, which enter the cerebellum via the middle cerebellar peduncle (MCP), can be used as CS (Steinmetz et al 1989; Steinmetz 1990).

Although the evidence is very strong that the cerebellum is important for classical conditioning, it is still not clear precisely which aspects of conditioning are controlled by the cerebellum. It is possible that the CS information undergoes im-

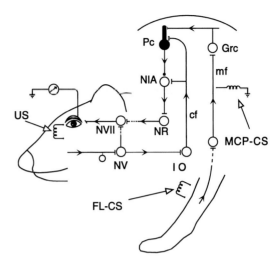

Figure 1: Experimental setup with hypothetical wiring diagram. Conditioned stimulation was applied to a forelimb (FL–CS) or to the middle cerebellar peduncle (MCP–CS). The unconditioned stimulus (US) was applied to the periorbital area. The CR was recorded as EMG activity from the upper eyelid. The hypothetical pathway for the CS is from the forelimb via mossy fibers (mf), granule cells (Grc), to the Purkinje cell (Pc). The hypothetical pathway for the US from the periorbital area is via the trigeminal nucleus (NV), the inferior olive (IO), and climbing fibers (cf) to Pc. Output is from the cerebellar cortex to the anterior interpositus nucleus (NIA), red nucleus (NR), and facial nucleus (NVII) to the orbicularis oculi muscle.

portant processing in the brain stem or in forebrain structures before it reaches the cerebellum. For instance, several investigators have suggested that the hippocampus performs a temporal processing that is utilized by the cerebellum (see e.g., Port et al. 1986; Christiansen and Schmajuk 1992). It is therefore of interest to determine to what extent conditioning, which uses direct stimulation of the mossy fiber input to the cerebellum as the CS, resembles conditioning to a peripheral CS.

In the first part of the study, we investigated the effects on CR latency of changing the current intensity or pulse frequency of a CS consisting of a train of electrical stimuli to the forelimb. In the second part, we studied the effects of changing the frequency of the CS when using direct stimulation of mossy fibers in the MCP as the CS. If the effects of varying the intensity or train frequency of a peripheral CS can be mimicked by variations in the frequency of direct mossy fiber stimulation, this would support the hypothesis that conditioning occurs in the cerebellum and that the CS is transmitted via mossy fibers. It would also indicate that the essential parts of the timing mechanisms reside in the cerebellum.

Materials and Methods

ANESTHESIA AND SURGERY

The experiments were performed on 23 male ferrets (0.8–1.8 kg). The animals were deeply anesthetised with isoflurane (Abbot Laboratories Ltd., UK, 1.5%–2% in a mixture of O_2 and N_2O). They were initially placed in a box into which anesthetic gas was directed. When deep anesthesia had been achieved, a tracheotomy was performed and the gas was then channeled directly into a tracheal tube. The level of anesthesia was monitored regularly by testing withdrawal reflexes.

The animal's head was fixed to a stereotaxic frame. The skull was opened on the left side, and the caudal half of the right and the left cerebral hemispheres and a substantial part of the thalamus on the left side were removed by aspiration. The aspiration exposed the left lateral part of the cerebellum and the superior and the inferior colliculus of the brain stem. The animal was then decerebrated by a section with a blunt spatula through the brain stem 0.5–1 mm rostral to the superior colliculus and the red nucleus. The completeness of the decerebration was verified in all cases by postmortem examination. In those animals in which stimulation of MCP was tested, the peduncle was exposed on the left side by removing the most lateral part of the tentorium. This procedure also exposed a small portion of the trigeminal nerve. To further expose the MCP, in four animals, the most lateral part of the inferior colliculus was aspirated. Bleeding was controlled with gelfoam. After decerebration the anesthesia was terminated. The end-expiratory CO_2 concentration, arterial blood pressure, and the rectal temperature were monitored continuously and kept within physiological limits. Throughout the experiment the animals received artificial ventilation and a continuous intravenous infusion (50 mg/ml of glucose, isotonic acetate Ringer's solution, 60 mg/ml of Macrodex with NaCl; proportion 1:1:1; 1 ml/kg per hour).

TRAINING PROCEDURES

All of the animals were first trained with a CS consisting of a 300-msec train of electrical stimuli to the left forelimb (FL-CS; 0.2 msec square wave pulses, 50 Hz, 1 mA) applied through two subcutaneous needle electrodes, one placed laterally and the other medially on the left upper forelimb. Later in the experiments, changes were introduced in the CS parameters, which are described in Results. In six animals the forelimb CS was replaced during the experiment by direct stimulation of mossy fibers using a tungsten wire electrode (diameter and length of deinsulated tip were 50 and 75 µm, respectively) inserted into the MCP (MCP-CS; 0.1 msec square wave pulses; other parameters are described in Results). The MCP had been exposed during the surgery, and the electrode could be positioned under visual guidance on the surface of the peduncle where it enters the cerebellum. The electrode was then lowered vertically 0.5–1 mm. The depth of the electrode tip was adjusted so that CRs could be reliably elicited by MCP stimulation. The strength required for this varied among animals (18–100 µA). If the electrode was lowered too deep (1–3 mm), the trigeminal nerve was activated. As shown in Figure 6D (below), single-pulse stimulation of the trigeminal nerve at 10 µA elicited short latency eyelid electromyograph (EMG) responses that were very similar in latency to responses elicited by the US (~4 msec). The stimulation site in the MCP was always verified by histological examination. The formaline-perfused cerebellum was sectioned and stained with Cresyl violet.

The US consisted of periorbital electrical stimulation through two stainless steel electrodes, the tips of which were inserted into the skin of the medial part of the periorbital area, ~5 mm apart. Three square pulses of 0.5 msec duration (50 Hz) were used. The stimulus strength was 3 mA. The excitability of the eye-blink reflex pathway was tested in naive animals or during adaptation by stimulating the periorbital area with a low-intensity US (one pulse of 0.5 msec and 0.5–1 mA) without a preceding CS.

The US was applied at the termination of the CS, so the interstimulus interval (ISI) was 300 msec. CRs were obtained in all animals. In previous work with this preparation in our laboratory, neither sensitization nor pseudoconditioning has ever been observed (Hesslow and Ivarsson 1996; Ivarsson et al. 1997).

The intertrial interval (ITI) was kept constant at 20 sec throughout most of the experiment. To exclude the possibility that the responses acquired during training were attributable to temporal conditioning, the ITI was occasionally increased to 30-60 sec in a pseudorandom manner over at least 15 trials. The acquired CRs always remained time-locked to the CS. This test was performed in every animal.

The eye-blink responses were monitored by EMG recordings from the orbicularis oculi muscles through two stainless steel electrodes inserted into the eyelid, ~3-5 mm above the lateral margin of the left eye. Electrophysiological recordings and analysis were performed with computer software developed in our laboratory. The sampling rate was 5 kHz. To reduce the baseline shift caused by the US, the recordings were run through a digital high-pass filter before analysis began.

CALCULATIONS OF CR MAGNITUDE, UR MAGNITUDE, AND LATENCY TO CR ONSET AND PEAK

To analyze CR magnitude, UR magnitude, and CR latency, the EMG signal was rectified. The CR magnitude was defined as the integrated EMG activity from 60 to 299 msec (in CS-US trials) or to 700 msec (in CS-alone trials) after CS onset. The UR magnitude was the integrated EMG activity from 4 to 22 msec after US onset. The UR was analyzed only in trials where the US was a single pulse.

To standardize the measurements of onset and peak latency the EMG activity was integrated over 5-msec bins. The CR onset latency was taken as the start of the first 5-msec bin, of which the EMG activity exceeded the spontaneous activity during 100 msec prior to CS onset by 200%. The latency to peak was the latency to the start of the bin with the highest amplitude. An example of CR onset and peak in one trial is indicated in Figure 6B (below). To be classified as a CR the EMG response had to begin between 60 msec and 299 msec after the CS onset and have a CR peak amplitude of at least 200% of the spontaneous activity.

Results

EFFECT OF VARYING THE FL-CS INTENSITY ON CR LATENCY

Animals were trained until stable conditioning had been achieved. This took 5-8 hr (500-1000 trials) at which time CRs were elicited on at least 95% of the trials. The onset latency of the CRs varied among the animals; the range of mean onset latencies was 99-248 msec. The latencies were relatively stable in each individual animal; the s.D. of the CR onset latency was <30 msec in 20 of the 23 animals, 35 msec in two animals, and 41 msec in one animal.

When the animals had acquired stable CRs, the CS intensity was shifted from the training level of 1 to 2 mA. This caused an immediate shortening of the onset latency [observed in 23 of 23 animals tested; (23/23)]. The size of this effect varied among animals. The mean latency change was 59 msec, and the range was 15-107 msec. The onset latency remained at the shortened level until the CS was reset to the original 1 mA after 10-24 trials. Results from a representative experiment are shown in Figure 2A. A second test in 13 of these animals had the same effect.

Increasing the CS intensity also caused an increase in CR magnitude (Fig. 2C). The shortening of the onset latency might therefore simply be a reflection of an increased CR amplitude. In that case, one would not expect any change in the latency to the peak amplitude of the CR. However, the latency to peak was also shortened by increasing CS intensity (20/23; Fig. 2B) by about the same amount as the onset latency. The mean change in latency to peak in the 20 animals varied between 24 and 96 msec as compared to 17 and 107 msec for the onset latency change. This point is illustrated further in Figure 2E, which shows representative examples of two smoothed averages of 10 rectified CRs each, elicited by 1- and 2-mA CSs, respectively. Both the onset latency and the latency to peak of the CR elicited by the stronger CS are shortened. The records were taken from CS-alone trials to get a clear picture of the topography of the CRs. It is noteworthy that the increased magnitude of these CRs was mainly attributable to a longer duration.

The change in CR latency was immediate and did not require any training. This was illustrated by experiments in which the CS intensity was alternated from trial to trial between 1 and 2 mA. In these experiments, the latencies also changed in a regular manner from trial to trial (11/11). An example is shown in Figure 2F, which shows records from five consecutive trials.

It is well known that when the ISI is changed, the latency to the peak of the CR no longer coincides with the onset of the US. After further train-

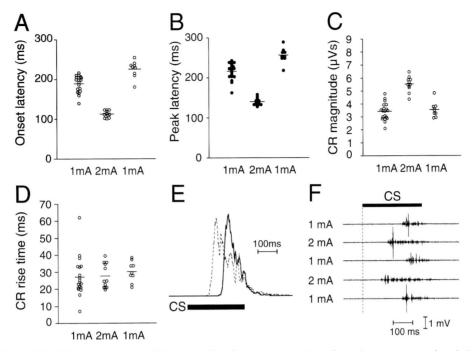

Figure 2: Effect of FL–CS intensity on the CR latency. The diagrams in *A–D* are from the same animal and show the effects on onset latency (*A*), latency to peak (*B*), CR magnitude (*C*), and CR rise time (*D*) when the CS intensity was 1 mA, increased to 2 mA, and then reset to 1 mA, respectively. Each circle corresponds to one trial, and the horizontal bar is the corresponding mean. The effect on CR topography is illustrated in *E* by rectified and averaged EMG recordings from CS-alone trials with a stimulation intensity of first 1 mA (solid line, $n = 10$) and then 2 mA (broken line, $n = 11$). These records were smoothed by applying a running average of 19 values, corresponding to 3.8 msec. (*F*) Five consecutive EMG records from the upper eyelid in another animal when the CS intensity was alternated from trial to trial.

ing the CR latency changes until the peak again coincides with the postshift US onset. Because increased CS intensity also causes a suboptimal CR (with latency of the CR too short to give maximal protection) it may be asked if this suboptimal CR, with further training, will return to the original latency. This suggestion is supported by an observation made in some experiments, namely that when the CS intensity was reset to the original 1 mA, after 10 trials or more at high CS intensity, the CR latencies were actually slightly longer than during the first control period (6/10). An example can be seen in Figure 2, A and B. The slightly longer latency may indicate an adaptation to the high CS intensity.

Accordingly, adaptability of CR latency was tested in 10 animals that had previously shown a latency shift on increased CS intensity, by increasing the CS intensity from 1 to 2 mA and then maintaining it at 2 mA. The increase in CS intensity from 1 to 2 mA immediately induced a decrease in CR onset latency and latency to peak, as described above. As the CS intensity was maintained at 2 mA,

however, both the onset and peak latencies were gradually prolonged and eventually returned to the control level. To estimate the rate of this adaptation, we divided the training period into five trial sessions and calculated the average latencies during each session. The number of sessions required until the onset latencies had increased to at least 95% of the average latency during a preceding control period varied between 4 and 22 (20 and 110 trials). We chose to illustrate this with plots from two extreme cases, with the one in Figure 3A being an animal with fast adaptation, and in Figure 3B an animal with slow adaptation of the CR latency. A second and third test in one of these animals had the same effect. The results are summarized in Table 1.

EFFECT OF VARYING FL–CS TRAIN FREQUENCY ON CR LATENCY

When the intensity of natural skin stimulation is increased, the number of activated afferent fibers

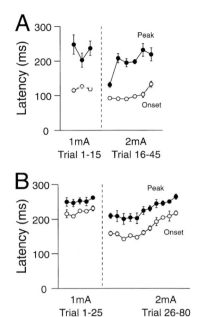

Figure 3: Adaptation of CR latency to high FL–CS intensity. When CS intensity was increased from 1 to 2 mA, the CR onset latency (○) and latency to peak (●) were immediately shortened. When the CS intensity was maintained at 2 mA, the onset latency and the latency to peak gradually returned to control levels. (*A*) A case with a fast return; (*B*) a case with a slow return to control latency. Each data point represents the mean ±S.E.M of five consecutive trials.

and also the impulse frequency in these fibers will usually increase. As increasing the intensity of electrical skin stimulation decreases CR latency, it may be expected that changing the frequency of such stimulation would be comparable to changing the intensity and hence have the same effect on CR latency. Previous studies of the effect of CS frequency on the latency of eye-blink CRs have only used tone as the CS. But changing the frequency of the tone CS is not comparable to changing the intensity and cannot be compared to varying the frequency of electrical skin stimulation. Different tones are coded in the cochlea by different, although overlapping, populations of hair cells and may be viewed as different CSs. Changes in CS frequency can then be interpreted as a test of generalization between two closely related CSs. This is indicated by experiments where test trials with increased CS tone frequency decreased the likelihood of a CR but not the CR latency (Kehoe and Napier 1991).

We tested the effect of increasing the frequency of the CS train from 50 to 100 Hz in 13 animals. The effect was very similar to that of increasing the CS intensity (Fig. 4). Both the CR onset and peak latencies were shortened by increased CS train frequency (13/13). The mean change of latency to onset was 67 msec and varied between animals from 21 to 123 msec. The mean change in peak latency was 40 msec and varied from 12 to 94 msec. The test was repeated in five of these animals and gave the same result. The effect of CS train frequency was also tested by alternating between high and low train frequencies from trial to trial. This resulted in immediate changes in CR latencies (12/12).

When the CS train frequency was maintained at 100 Hz in four animals, the CR latencies gradually returned to at least 95% of the preshift latency within 2–21 sessions (10–105 trials). This is illustrated with results from two animals in Figure 4, C and D. A second and third test in one of these animals had the same effect. The results are summarized in Table 1.

EFFECT OF VARYING FL–CS INTENSITY AND TRAIN FREQUENCY ON UNCONDITIONED RESPONSE MAGNITUDE

It might be claimed that the decrease in CR latency was a result of an increase in the general excitability or arousal induced by the increased intensity, and not specific to the CS–CR pathway. To test this possibility, the effect of increasing CS intensity on eye-blink reflex excitability was studied in naive, nonconditioned animals. Three naive, nontrained animals received first US alone trials (one pulse of 1 msec; 0.5 mA). Then the animals received paired CS–US stimulation but with a weak US (one pulse of 1 msec; 0.5 mA).

The weak US elicited submaximal unconditioned responses (URs), smaller than those elicited by 3 mA. Preceding the US with a weak (1 mA) or a strong (2 mA) CS had no significant effect on the amplitude of the UR (Fig. 5A). This suggests that the CR latency decrease was not attributable to a general increase in excitability or arousal.

A similar test was performed to exclude that the adaptation of the CR latency to a permanent high-intensity CS was attributable to a change in excitability. Thus, UR magnitude was measured before, during, and after such adaptation, again with no systematic effects on excitability (Fig. 5B).

Equivalent tests of the effects of CS parameters on UR magnitude were performed in three animals

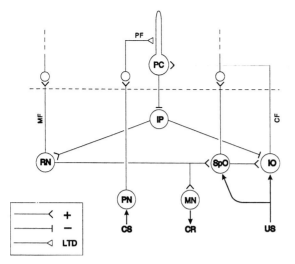

Figure 1: Cerebellar and brain stem circuits underlying eye-blink conditioning (after Rosenfield and Moore 1995). (MF) mossy fibers; (PF) parallel fibers; (PC) Purkinje cell; (RN) red nucleus; (IP) interpositus nucleus; (SpO) spinal trigeminal nucleus pars oralis; (CF) climbing fiber; (IO) inferior olivary nucleus; (PN) pontine nucleus; (MN) motoneurons; (LTD) long-term depression; (CS) conditioned stimulus; (CR) conditioned response; (US) unconditioned stimulus.

ror CRs as they unfold in time, encoding their latency and peak amplitude. How does this come about?

There are several ways to approach this question. The one we have favored relies on the development of computational models that can be expressed as neural networks. This involves three steps: (1) Devise real-time computational models that describe as much of the known behavioral and physiological evidence as possible; (2) devise an implementation scheme that aligns features of the model with involved neural circuits; and (3) test implications of the model and its implementation in novel experiments.

This article applies these steps to the Time Derivative (TD) Model of Pavlovian Reinforcement (Sutton and Barto 1990). Before presenting the TD model and a plausible implementation within the cerebellum, we review related computational models and their limitations. Next, we present the TD model and illustrate its predictions in simple and complex paradigms. Simple paradigms are those that involve training with a single conditioned stimulus (CS) presented repeatedly with an unconditioned stimulus (US) at a fixed CS-US interval (interstimulus interval, or ISI). Complex paradigms are those involving integration of information from two or more types of training trials. Familiar examples of complex paradigms include Kamin blocking, conditioned inhibition, mixed CS-US interval training (temporal uncertainty), and higher-order conditioning.

Theoretical Background

Contemporary computational models of learning in classical conditioning trace their origins to the Rescorla-Wagner (RW) model (Rescorla and Wagner 1972). The RW model supplanted earlier mathematical models, such as the statistical learning theory, because of its ability to generate accurate predictions in a number of complex conditioning paradigms, including Kamin blocking and conditioned inhibition. The RW model can be expressed by an equation that specifies how the associative connection (V_i) between the ith CS and the CR is modified from one trial ($t - 1$) to the next (t) as a function of pairing with a US.

$$\Delta V_i(t) = \beta[\lambda(t) - Y(t-1)] \times \alpha X_i(t) \quad (1)$$

where

$$Y(t) = \sum_j V_j(t) X_j(t) \quad (2)$$

The subscript j indexes all CSs that are active on trial t. α and β are rate parameters ($0 < \alpha, \beta \leq 1$), $\lambda(t)$ is the strength of the US, and $X_i(t)$ denotes the presence [$X_i(t) = 1$] or the absence [$X_i(t) = 0$] of CS_i.

The RW model features a "competitive" learning rule. This means that connection weights are adjusted so that the sum of the values of all CSs present on a trial approach the strength of the US, λ. Kamin blocking occurs whenever a new CS is combined with one that had been previously trained. Blocking occurs because the previously trained CS fully predicts the US, having garnered most of the associative value the US can support. Conditioned inhibition occurs whenever a CS is combined with a previously trained excitatory CS but not paired with the US. Both CSs lose associative value, but because the added CS is never paired with the US, its value becomes increasingly negative. The net effect of combining an excitatory and inhibitory CS after training in this paradigm is a marked suppression or cancellation of the CR.

Although the RW model and other trial-level models continue to impact learning theory in many domains, their limitations motivated the emergence of real-time models of classical conditioning (Klopf 1988). Two limitations of trial-level models are their inability to generate CS–US interval functions or higher-order conditioning (Sutton and Barto 1990). Real-time models invoke other processes to account for these phenomena. In real-time models, the timing and duration of CSs and the US determine how learning unfolds. Real-time models imply that time is treated as a continuous variable and that learning can therefore be expressed by differential equations in time. In practice (and for simulation purposes), real-time models segment time into a series of computational epochs with a fine enough grain to capture dynamical events on a time scale appropriate for the preparation.

Sutton and Barto's (SB) Time Derivative (TD) model was the first real-time model to generate CS–US interval functions and higher-order conditioning (Sutton and Barto 1981; Moore et al. 1986). The SB model is a member of a class of models that Sutton and Barto (1990) have referred to as $\dot{Y}$ theories of learning. These theories adjust connection weights (associative values) according to the first time derivative of the learning system's output or response, $Y(t)$. If the output on the current time step is greater than that of the preceding time step, active connections are strengthened. If the output on the current time step is less than that of the preceding time step, active connections are weakened. The US affects learning only to the extent that it contributes to output.

$$\Delta V_i = \beta \dot{Y} \times \alpha_i \overline{X}_i \quad (3)$$

where

$$\dot{Y} = Y(t) - Y(t-1) \quad (4)$$

Any device that would implement $\dot{Y}$ learning must receive information about $Y(t)$ and $Y(t-1)$.

The following three equations specify the SB model.

$$\Delta V_i(t) = \beta[Y(t) - Y(t-1)] \times \alpha \overline{X}_i(t) \quad (5)$$

where

$$Y(t) = \sum_j V_j(t) X_j(t) + \lambda(t) \quad (6)$$

$\overline{X}_i(t)$ specifies the *eligibility* of the *i*th connection for modificaiton. A CS's eligibility decreases geometrically after offset according to the following equation:

$$\overline{X}_i(t+1) = \overline{X}_i(t) + \delta[X_i(t) - \overline{X}_i(t)] \quad (7)$$

A slowly decaying eligibility allows modification of connection weights for CSs that are no longer contributing to output.

Although the SB model provides a framework for describing aspects of eye-blink conditioning that the RW model cannot address, it does not generate realistic-appearing CRs. There are two problems with the SB model in this regard. First, the model does not capture the fact that the onset of CRs is delayed with respect to CS onset and peaks at the time of the US. Second, the model predicts that response amplitude, $Y(t)$, in the presence of the US is too large (equation 6). For example, if CS elicits a 10-mm eyelid closure after training, and if the US presented alone elicits a 10-mm eyelid closure, then the combination of the CS and US should elicit a 20-mm response. This does not happen because eyelids can close only so far and no further. Moore et al. (1986) devised a variant of the SB model that attempted to solve these problems. First, it was assumed that the onset of a CS does not trigger an immediate response but instead causes the gradual rise in amplitude that peaks at a fixed time and that is sustained until the CS is withdrawn. Second, equation 6 was modified so that the contribution of the US, λ, to $Y(t)$ decreases progressively as V_i increases over training.

Despite these modifications, the SB model does not provide an adequate model for generating realistic CRs (Desmond 1990). For one thing, the SB model cannot describe the shifts in CR timing that occur when the CS–US interval is changed during training. Nor can it describe the fact that CRs can be elicited by stimulus offsets as well as stimulus onsets (Desmond and Moore 1991b). These problems with the SB model led to the development of another model, similar to SB, but with additional features that overcame the former's limitations (Desmond and Moore 1988; 1991b; Moore et al. 1989; Moore 1992; Moore and Desmond 1992). We refer to this model as VET, for associative values based on expectations about timing.

The VET model assumes that both stimulus onsets and offsets cause cascades of spreading activation. This spreading activation sequentially engages time-tagged "input elements". These input elements can be regarded as serial components of the nominal CS (Sutton and Barto 1990).[2] Computations on these sequentially activated elements give rise to CRs that mirror the CS–US intervals employed in training. Such CRs are said to be *temporally adaptive*. The timing structure of the VET model can be applied to the TD model, as illustrated later on.

Although it is an advance on the SB model, the VET model has one important deficiency. Like the RW model (but unlike the SB model), it cannot generate higher-order conditioning. That is, it does not allow for the establishment of a CR when a novel stimulus is paired with a previously trained CS. The SB model can generate higher-order conditioning because the strength of conditioning (i.e., the value or weight of the connection between CSs and the CR) depends only on the changes in output from one time step to the next. Increases in output result in increases in connection weight; decreases in output result in decreases in connection weight. The US is not essential for this learning. In contrast, changes in connection weights in VET (and RW model) revolve about the US. Weights increase when the US is presented and decrease when it is omitted, subject to the constraints of competitive learning.

Hybrid Learning: TD Model

Experience with the SB and VET models suggested the need for a model that combines the best features of both. Sutton and Barto (1990) showed that the TD model overcomes the limitations of the SB and VET models. We have extended the TD model by incorporating the timing structure of the VET model. The TD learning rule is given by the following equation:

$$\Delta V_i(t) = \beta[\lambda(t) + \gamma Y(t) - Y(t-1)] \times \alpha \overline{X}_i(t) \quad (8)$$

[2]Sutton and Barto (1990) refer to this timing structure as a complete serial compound (CSC). The TD model with the CSC representation of time has been applied to predictive timing and error correction in the dopaminergic reward system of monkeys (Schultz et al. 1997).

where

$$Y(t) = \sum_j V_j(t) X_j(t) \quad (9)$$

As in the SB model, $\overline{X}_i$ refers to eligibility for modification.

$$\overline{X}_i(t+1) = \overline{X}_i(t) + \delta[X_i(t) - \overline{X}_i(t)] \quad (10)$$

The hybrid nature of the learning rule is evident in the fact that changes in connection weights depend on two reinforcement components. One component is contributed by the US, λ. The other component is contributed by a $\dot{Y}$ component, $\gamma Y(t) - Y(t-1)$.

A key feature of the TD model is the parameter γ ($0 < \gamma \leq 1$). γ is referred to as the *discount* parameter because $Y(t)$ is not known with certainty until after the fact. That is, $Y(t)$ must be estimated by using the connection weights computed on the previous time step, $Y(t) = \Sigma_j V_j(t-1) \times X_j(t)$. γ can be regarded as the penalty for using $V(t-1)$ as an estimate of $V(t)$.

The TD model overcomes the deficiencies of earlier models while retaining their ability to describe complex paradigms such as Kamin blocking and conditioned inhibition. However, for the TD model to encompass CR timing and topography, Sutton and Barto (1990) proposed that the elapsed time between the onset of a CS and the US be segmented into an ordered sequence of serial components. These serial components are, for all intents and purposes, the same as the time-tagged input elements of the VET model. The subscript i in the equations of the TD model refers to a single serial component.

Figure 2 shows simulated CRs with the TD model with variations of two parameters, γ, and δ (λ held constant). Both parameters contribute to the latency and amplitude of CRs. These simulated CRs are realistic in that they resemble goal gradients: Amplitude rises progressively to peak at the time of US onset.

For the TD model to describe CR timing and topography in trace conditioning and in complex paradigms involving multiple CSs, we adopt the timing structure of the VET model. Both CS onsets and offsets are assumed to trigger cascades of spreading activation. This spreading activation is mapped onto the serial components of the TD model. Each nominal CS, such as a tone and a light,

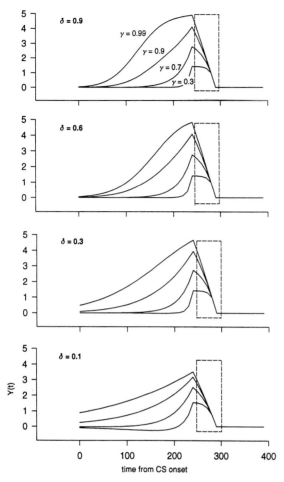

Figure 2: Simulated CRs, $Y(t)$, after 200 trials as a function of γ and δ. Time steps in this and other simulations are 10 msec, $\alpha = 0.05$, $\beta = 1.0$, and $\lambda = 1.0$. The rectangle in each panel indicates the duration (50 msec) and intensity of the US [scales in terms of $Y(t)$]. Note that CR timing is determined primarily by the discount factor, γ.

initiates an independent cascade that sequentially activates the variables X_i in the model ($i = 1,2,3,...$). The duration of activation of a serial component need not be fixed or constant, but for simulation purposes we have assumed a temporal grain of 10 msec. Hence, this is the assumed duration of activation of a serial component. When activated, $X_i = 1$; when inactivated (after 10 msec), X_i resets to a baseline of 0 as the next serial component, X_{i+1} is activated. Although X_i is no longer active and therefore no longer contributes to Y, the output or response, its connection to the output, V_i, remains eligible for modification over succeeding time steps. Eligibility decays at a rate determined by δ. And, as mentioned, just as a nominal CS initiates a cascade of activation among serial

components, so too does its offset. The two cascades are assumed to operate independently and in parallel. There are limits on how long these cascades might last, that is, on the number of sequentially activated elements in each cascade. The only requirement is that these cascades span the CS–US intervals employed in training.

TD Model and Temporal Uncertainty

Like the VET model, the TD model predicts the timing and amplitude of CR waveforms in complex training paradigms. One such paradigm involves training with a random mixture of two CS–US intervals. We refer to this training paradigm as conditioning under temporal uncertainty (Millenson et al. 1977). If the two CS–US intervals are sufficiently different (e.g., 300 vs. 700 msec), then rabbits learn to generate bimodal CRs with amplitude peaks at the temporal loci of the two times of US occurrence.

Figure 3 shows simulated bimodal CRs following temporal uncertainty training with CS–US intervals of 300 and 700 msec. In Figure 3A, the CS duration is 300 msec. The simulated bimodal CR to a CS-alone probe trial has two amplitude peaks. The one at 700 msec is larger than the one at 300 msec because CS offset contributes to the 700-msec peak but not the 300-msec peak. In Figure 3B, the CS duration is 800 msec. Because CS offset occurs after the longest CS–US interval, its cascade does not contribute to the second peak. Consequently, the two peaks have the same amplitude.

Uncertainty Training and the Cerebellum

How is experience with temporal uncertainty represented in the cerebellum? Do firing patterns of single neurons express the timing and amplitude eye-blink CRs in a manner predicted by the TD model? We have begun to address this question in experiments with rabbits, employing the temporal uncertainty protocol shown in Figure 4A (J.-S. Choi and J.W. Moore unpubl.). The CS is a 300-msec tone. The US is a mild electric current applied to the periocular tissue of the right eye. Training consists of a random mixture of two trial types. On trial-type 1, the CS–US interval is 300 msec. On trial-type 2, the CS–US interval is 700 msec. After 20 daily sessions (80 trials/session with an average

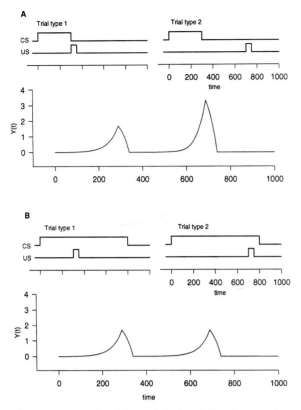

Figure 3: Simulated bimodal CRs following training with a random mix of two CS–US intervals. (*A*) Two types of training trials and a simulated bimodal CR waveform. The CS is 300 msec in duration. The CS–US interval is 300 msec on trial type 1 and 700 msec on trial type 2. Notice that the second CR peak is larger than the first CR peak. (*B*) Two types of training trials and a simulated CR waveform. The CS is 800 msec in duration. The CS–US intervals are the same as in *A*. Note that the two CR peaks have the same amplitude.

intertrial interval of 25 sec), rabbits are surgically prepared for microelectrode recording (Berthier and Moore 1990). Training resumes follow recovery to ensure that bimodal CRs are well-established. A microelectrode is then advanced through the cerebellar cortex and into deep nucleus IP. During recording, the two trial types employed in training continue to be presented, but there are also CS-alone probe trials.

Figure 4B shows a single CS-alone probe trial (CS onset occurs at time = 350 msec). The top trace is a record of eyelid position as a function of time. Notice that there are two amplitude peaks and that these are located at the loci of the US. The second peak is larger than the first, in agreement with the TD model as shown in Figure 3A. The second trace shows the firing of an IP neuron on this trial. Note that the rate and duration of firing are highly related to the two CR peaks. Figure 4C shows the averaged CR topography (top trace) and spike histogram for all probe trials with this neuron.

Figure 4, D and E, shows averaged CR topographies and spike histograms for this neuron on reinforced trials (upward arrows mark the US). Figure 4D is interesting because it shows that the occurrence of the US cancels the second amplitude peak and terminates the CS-triggered spiking. This is an important observation because it suggests (but does not prove) that the US acts as a conditioned inhibitor.[3] In terms of the model, the US initiates a cascade of activation of serial components that is never paired with the US. In training trials on which the US occurs at 300 msec (trial type 1), the US-triggered cascade exists alongside two CS-triggered cascades, an onset cascade and an offset cascade. The CS-triggered cascades are paired with the US on trial type 2 (700-msec CS–US interval), so they are excitatory. In contrast, because there is only one US per trial, the US-triggered cascade is never paired with the US. According to competitive learning rules, serial components of the cascade triggered by the US become conditioned inhibitors and therefore they have the capacity to suppress CRs.

If this scenario is correct, it indicates that conditioned inhibition is expressed at the level of single cerebellar neurons. The expression of both excitation and inhibition within the same cerebellar neuron would be an important discovery about the locus of action of conditioned inhibition (Blazis and Moore 1991). Such an observation would be consistent with evidence that the cerebellum is the locus of extinction, the gradual decline of the CR through repeated presentations of a CS without reinforcement: (1) Ramnani and Yeo (1996) showed that reversible inactivation of nucleus IP protects against extinction; (2) Perrett and Mauk (1995) showed that vermal cerebellar lesions interfere with extinction, just as lesions of HVI interfere with CR acquisition and performance.

Several investigators have suggested that cerebellar PCs mediate the CR-related activity ob-

---

[3]It is unlikely that the US simply terminates all timing cascades, thereby accounting for the absence of the second amplitude peak on 300-msec probe trials. Instead, this capacity develops progressively with training; the US does not terminate the second peak until training is well advanced.

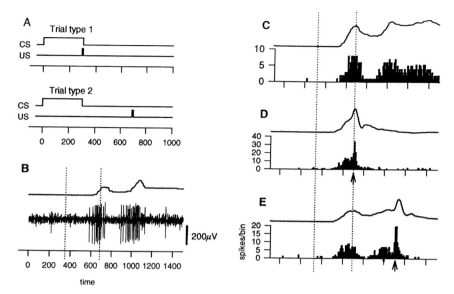

Figure 4: Firing patterns of an IP neuron related to bimodal CRs. (*A*) Two types of training trials: CS duration = 300 msec. CS–US intervals are 300 msec (trial type 1) and 700 msec (trial type 2). (*B*) A single CS-alone trial: Top trace shows CR waveform; second trace shows neuronal response. Vertical dotted lines mark CS onset and offset. (*C*) Average CR waveforms and spike histogram on probe trials (*n* = 6). (*D*) Average response waveform and spike histogram on type 1 reinforced trials (*n* = 11). Arrows mark the US. (*E*) Average response waveform and spike histogram on type 2 reinforced trials (*n* = 8).

served in nucleus IP (Berthier and Moore 1986; Thompson 1986).[4] Because PCs are inhibitory (Fig. 1), they can generate the CR-related increases in firing in IP neurons only by decreasing their rate of firing. Figure 5, which summarizes results from a PC in a different animal than the one in Figure 4, indicates that this may be the case. Figure 5A shows a single probe trial following temporal uncertainty training in the protocol of Figure 4. The top trace shows two CR peaks, with the second peak larger than the first. The second trace shows pauses in simple spike rates related to each CR peak. Figure 5, shows the averaged CR waveforms and spike histogram for this cell on probe trials. Figure 5, C and D, shows averaged CR waveforms and spike histograms for this cell on the two reinforced trial types. The histograms show that the rate of simple spiking decreased in anticipation of each of the two amplitude peaks. The US caused a brief burst of activity that was followed by a post-US pause on the order of 100 msec. There is some indication of conditioned inhibition by the US in Figure 5C, because of the transitory increase in spiking 100 msec after the US.[5]

Cerebellar Implementation of the TD Model

Our recording studies indicate that the full complexity of conditioned eye blinks in a temporal uncertainty paradigm can be represented in the firing of single IP neurons. Furthermore, it is possible that this complexity is also captured in the activity of individual PCs (Fig. 5), as suggested by theorists (e.g., Moore et al. 1989; Fiala et al. 1996). From this perspective, IP neurons execute motor programs by inverting signals generated by PCs. We turn next to a consideration of how the TD model's learning rule might be implemented in the cerebellar cortex.

TD learning can be implemented in the cerebellum by aligning known anatomical ingredients with elements of the learning rule. In TD learning,

---

[4]The activity of a PC expressing long-term depression (LTD) would not to be as highly correlated with eye-blink CRs as would the activity of an IP neuron, because of the many-to-one convergence of PCs onto IP neurons (Ito 1984).

[5]Conditioned inhibition would be implemented by parallel fiber (PF)/PC synapses expressing long-term potentiation (LTP). It is not presently known whether LTP and LTD synapses can coexist on the same PC, as suggested by Fig. 5.

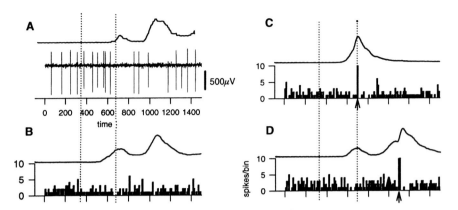

Figure 5: Firing patterns of a PC related to bimodal CRs. (*A*) A single CS-alone probe trial: Top trace shows CR waveform; second trace shows neuronal response. Vertical dotted lines mark CS onset and offset. (*B*) Average CR waveforms and spike histogram on probe trials (*n* = 13). (*C*) Average response waveform and spike histogram on type 1 reinforced trials (*n* = 11). (*D*) Average response waveform and spike histogram on type 2 reinforced trials (*n* = 14).

we assume that each computational time step after the onset or offset of a CS is represented by anatomically distinct inputs to the cerebellum. The onset or offset of a CS initiates a spreading pattern of activation among neurons tied to whatever sense modality is involved. This spreading of activation, possibly under entrainment from oscillators, engages pontine nuclear (PN) cells, the primary source of cerebellar mossy fibers (MFs), and their associated granule cells. Therefore, timing elements should be regarded as ensembles that include PN cells, MFs, granule cells, parallel fibers and influences from intrinsic cerebellar neurons such as Golgi cells. This may be why CR timing is disrupted by lesions of the cerebellar cortex (Perrett et al. 1993). Entrainment by oscillators might occur at the level of the PN, as these are the nexus of neural influences from the lemniscal systems, midbrain, and forebrain (Wells et al. 1989). Fine-grain temporal segmentation might occur locally within the cerebellum, as proposed by Bullock et al. (1994). Coarse-grain temporal segmentation and coherence might occur globally with the participation of the hippocampus, as suggested by Grossberg and Merrill (1996).

The implementation relies on evidence from rabbit eye-blink conditioning that CR topography is formed in the cerebellar cortex through converging contiguous action of parallel fiber (PF) and climbing fiber (CF) input to PCs. This action produces synaptic long-term depression (LTD). Chen and Thompson (1995) and Schreurs et al. (1996) have demonstrated pairing-specific LTD of PCs in cerebellar slice preparations from rabbits, using parameters that support conditioning in intact animals. Consistent with the LTD hypothesis, Hesslow (1994) showed that stimulation of the cerebellar cortex (HVI) inhibits eye-blink CRs in decerebrate cats. Mechanisms of LTD in the cerebellum have been spelled out in recent articles (Kano et al. 1992; Konnerth et al. 1992; Schreurs and Alkon 1993; Eilers et al. 1995; Ghosh and Greenberg 1995; Hartell 1996; Schreurs et al. 1996, 1997; Kim and Thompson 1997).

Figure 1 incorporates recent anatomical findings by Rosenfield and Moore (1995) indicating the existence of projections to HVI from the RN and SpO. CS information ascends to granule cells in the cerebellar cortex (Larsell's lobule HVI) via MFs originating in the PN. Information about the US ascends to the cerebellar cortex by two routes, MF projections from the sensory SpO in Figure 1, and CF projections from the inferior olive (IO). A CR is generated within deep cerebellar nucleus IP, where the CR is formed by modulation from PCs. A full-blown CR is expressed as an increased rate of firing among IP neurons (e.g., Berthier and Moore 1990; Berthier et al. 1991). This activity is projected to the contralateral RN. From RN, activity is projected to motoneurons (MNs) that innervate the peripheral musculature controlling the position and movements of the eyelids and eyeball (Desmond and Moore 1991a). The RN also projects to SpO, giving rise to CR-related activity among these neurons (Richards et al. 1991).

Figure 1 depicts an inhibitory projection from IP to IO. The consequence of this arrangement is that olivary signals to PCs are suppressed when the

CR representation within IP is robust. This anatomical feature suggests that CFs are excited only when the US occurs *and* the CR is weak or absent. This scenario has been supported by Sears and Steinmetz (1991), who showed that neural activity recording within IO diminishes during CR acquisition. In the TD model, the CR is regarded as a prediction of US onset. Therefore, a CR should not occupy the same time step as the US. If it does, because its timing is somehow delayed or because its momentum extends beyond US onset, CF inputs to the cerebellar cortex would be inhibited by the projection from IP to IO. Trials on which this occurs would be tantamount to extinction trials, thereby countering the down-regulated state of PF/PC synapses that mediate CRs. PF/PC synapses expressing LTD would gradually lose this capacity until US-triggered CF inputs are no longer blocked by overly robust CRs. This feedback mechanism could be a vital aspect of CR timing, in that it would ensure that CRs do not become so robust that they lose precision in predicting US onsets.

The TD learning rule is not a simple competitive rule because of the $\gamma Y(t)$ term in equation 8. As noted previously, the TD learning rule is implemented by a combination of two reinforcement components. The first is donated by the US and represented by λ in equation 8. The implementation assumes that λ can be aligned with CF activation of PCs, which functions to produce LTD among coactive PF synapses, as depicted in Figure 1. The second reinforcement component is donated by the $\dot{Y}(t)$ terms in the learning rule, $\gamma Y(t) - Y(t-1)$. This information is conveyed to HVI by the projection from RN and SpO shown in Figure 1.

Figure 6 shows circuit elements, not shown in Figure 1, for implementing the $\dot{Y}(t)$ component of the learning rule. These components include the projections to the cerebellar cortex from the RN and SpO indicated in Figure 1. We hypothesize that the RN projection carries information (feedback) about $Y(t)$ to the cerebellar cortex as efference copy. PFs project this information to PCs that have collaterals to a set of Golgi cells. Because these projections are inhibitory (Ito 1984), these PCs invert the efference signal from the RN. In addition, the interpositioning of the PCs between the RN and Golgi cells attenuates the signal and implements the TD model's discount factor γ.

Because Golgi cells are inhibitory on granule cells, the effect of their inhibition by PCs receiving efference from the RN would be to disinhibit ac-

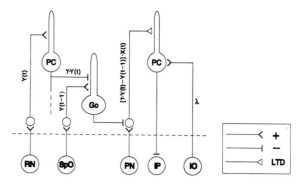

Figure 6: Neural circuits of the cerebellum implementing $\gamma Y(t)$ and other variables of the TD learning rule. (Go) Golgi cells.

tivity of granule cells. Because granule cells relay CS information from the PN to PCs involved in LTD and CR generation, disinhibition of granule cells by Golgi cells enhances the information flow from active CS components. Mathematically, the implementation assumes that the variables X_j in equation 9 engage granule cells. PFs arising from these granule cells trigger output, and they affect connection weights residing at PF/PC synapses in proportion to $Y(t) \times X_j$.

PCs driven by projections from the RN would increase their firing rate so as to mimic the representation of the CR as it passes through the RN enroute to MNs and SpO. Berthier and Moore (1986) recorded from several HVI PCs with CR-related increases in firing. Because increases in firing during a CS is inconsistent with the LTD hypothesis of CR generation, these PCs serve some other function. One possibility is that they inhibit motor programs incompatible with CR generation. Here, we are suggesting an additional function of these PCs, that of projecting inverted and discounted CR efference from the RN to Golgi cells.

The implementation assumes that the Golgi cells that receive the inverted efference from the RN also receive a direct, noninverted, excitatory projection from SpO. This projection carries information about the CR at time $t - \Delta t$. Therefore, the Golgi cell in Figure 6 fires at a rate determined by the differential between two inputs: $\gamma Y(t)$ donated by the RN and $Y(t - \Delta t)$ donated by SpO.[6] Hence,

---

[6]The PC that donates γ does not impose a significant delay in transmitting feedback from RN to Golgi cells. Conduction distances within the cerebellar cortex are on the order of tens of microns. Projections to the cerebellum from the brain stem, and from the RN to the SpO, are on the order of tens of millimeters.

Golgi cells act as $\dot{Y}(t)$ detectors. In terms of equation 8, $Y(t)$ is transmitted to cerebellar granule cells by the RN, and $Y(t-1)$ is transmitted to granule cells from SpO. The RN input engages PCs that inhibit Golgi cells responsible for gating inputs from CSs to PCs. Efference from SpO engages the same Golgi cells directly. Because Golgi cells are inhibitory on granule cells, the bigger the RN input relative to SpO input is, the bigger the signal from serial component CSs active at that time, be they from onset or offset cascades.

In this way, the Golgi cells that implement $\dot{Y}(t)$ reinforce and maintain the down-regulated state of active PF/PC synapses. PF/PC synapses that are activated by a CS element are down-regulated by the contiguous US-triggered activation of CF input from the IO. As CS elements earlier in the timing sequence become capable of evoking an output that anticipates the US, inhibition is relayed to the olive and the US loses its capacity to trigger a CF volley (see Fig. 1). However, the down-regulation of these synapses is maintained, and still earlier CS elements are recruited, by PFs carrying $\dot{Y} \times X_j$ to LTD-PCs, as indicated in Figure 6.[7]

In their recording study, Desmond and Moore (1991a) observed an average lead time of 36 msec between the initiation of CR-related firing in RN neurons and the peripherally observed CR. Richards et al. (1991) observed an average lead time of 20 msec in SpO neurons. Therefore, the time difference in CR-related efference arising from the two structures is on the order of 15-20 msec. This difference spans one 10-msec time step used in our simulations with the TD model. This temporal difference is consistent with a conduction velocity of 2 m/sec for the 10-mm trajectory of unmyelinated axons from the RN to rostral portions of the SpO. The 10-msec grain also ensures good resolution of fast transients. The fastest transients in eye-blink conditioning occur during unconditioned responses. At its fastest, the eyelids require 80 msec to close completely, with a peak velocity of 4-5 mm/20 msec.

---

[7]Hartell (1996) showed that strong PF activity can induce LTD at PF/PC synapses and that this depression extends to other, spatially separated synapses. Hartell (1996) also reported that LTD induced by strong PF stimulation occludes the form of LTD mediated by CFs. Schreurs et al. (1997) report a similar occlusion of CF-mediated LTD among PCs that express LTP. Hartell (1996) did not address the question of whether strong PF stimulation maintains and reinforces *previously established* CF-induced LTD, which is a core assumption of the implementation.

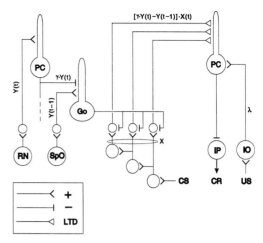

Figure 7: The complete TD implementation scheme showing three sequentially activated CS components, representing onset and offset cascades in the manner of the VET model of Desmond and Moore (1988). (Go) Golgi cells.

Efference from SpO neurons recorded among HVI PCs would tend to lag behind the peripherally observed CR, especially if it arises from more caudal portions of the structure. Berthier and Moore (1986) observed a continuum of lead and lag times among PCs that increased their firing to the CS. PCs that receive projections from the SpO (not shown in Fig. 6) would be expected to increase their firing but with a lag relative to those receiving projections from the RN. The proportion of CR-leading PCs observed by Berthier and Moore (1986) matched the number of CR-lagging PCs, which makes sense if the two populations reflect CR efference from two spatially separated sources, RN and SpO.

Figure 7 is an expanded version of Figure 6 showing three sets of granule cells associated with three serial component CSs. These components might arise from CS onset or offset. The degree to which information from any of these serial CS components reaches the PCs to which they project is determined by Golgi cells firing in proportion to $\dot{Y}(t)$, as just described. Figure 2 shows that TD-simulated CRs tend to be positively accelerating (contingent upon γ) up to the occurrence of the US, so $\dot{Y}(t)$ increases progressively over the CS-US interval. Therefore, those PF/PC synapses activated nearest the time of the CF signal from the US have the greatest impact in establishing and maintaining LTD. This mechanism ensures the appropriate form and timing of CRs.

Interpretations of $\dot{Y}$: Efference or Afference

Equation 8 emphasizes interpretations of $\dot{Y}$ as efference, but it is equally correct to interpret changes in associative values in terms of afference, substituting equation 9 into equation 8. Recent studies by Ramnani and Yeo (1996) and Hardiman et al. (1996) suggest that the efference interpretation of $\dot{Y}$ is correct. These studies show that temporary inactivation of IP by muscimol prevents extinction of the CR. That is, CS-alone trials that would normally lead to a gradual elimination of the CR instead had no effect whatsoever. When tested after the muscimol blockade had been removed, the previously established CR was at full strength but extinguished normally with continuing presentation of CS-alone trials. This finding is consistent with the efference interpretation of the TD implementation because inactivation of IP eliminates the CR and therefore prevents efference from the RN and SpO from reaching the putative site of learning in HVI. In terms of equation 8, connection weights cannot decrease if $Y(t)$ and $Y(t-1)$ are both equal to 0. If the afference interpretation were correct and efference plays no role in extinction, then inactivation of IP would not prevent extinction because afference arises from PN and bypasses IP enroute to HVI.

Implications of the Implementation

The implementation has several testable implications.

1. One implication is that some PCs decrease their firing rate in anticipation of CR peaks. These PCs express LTD. However, there are other PCs, perhaps the majority, that increase their rate of firing in relation to CR peaks, as reported by Berthier and Moore (1986). Some of these PCs express efference from the RN and SpO. Their function is to activate Golgi cells that modulate information flow through the granule cells. Another function would be to inhibit motor programs such as eye opening and saccadic movements that could interfere with eyelid closure.
2. The implementation specifies that Golgi cells that modulate the flow of CS information in granule cells fire in relation to *changes* in eyelid position, that is, they fire in relation to $\dot{Y}$. This property of Golgi cell-firing patterns has been reported by van Kan et al. (1993), in a study of monkey limb movements, and Edgley and Lidierth (1987), in a study of cat locomotion.
3. The implementation implies that reversible inactivation of the RN would prevent second-order conditioning. However, although inactivation of the RN would cause a temporary interruption of information flow that results in a CR, it would not prevent learning of the primary association between components of the CS and the US. This association proceeds with little disruption because the PN and the IO can still convey CS and US information to the cerebellar cortex. Evidence for this proposition comes from a study of rabbit eye-blink conditioning by Clark and Lavond (1993). They demonstrated that inactivation of the RN by cooling did not prevent learning, as CR magnitude recovered upon reactivation of the RN. However, inactivation of the RN would interrupt efference about the position of the eyelid at times t and $t - \Delta t$ from the RN and SpO. Thus, $\dot{Y}$ would not be available to the cerebellar cortex. According to the TD model, $\dot{Y}$ allows for increments of predictive associations in the absence of the US, as would occur in second-order conditioning. This being the case, inactivation of the RN would interfere with second-order conditioning. With the RN inactivated, animals trained with a mixture of first- and second-order (Kehoe et al. 1981) would be expected to show first-order learning, as in the Clark and Lavond (1993) study, but little or no second-order learning. Figure 8 shows a simulation of the failure of second-order conditioning by blocking the $Y(t)$ efference projection to HVI from the RN.[8]

Summary

We have described how the TD model of classical conditioning can generate appropriate CR waveforms in simple protocols and complex paradigms involving temporal uncertainty. We also suggest an implementation scheme for TD learning within the cerebellum. The implementation draws on neurobiological evidence regarding how LTD is

---

[8]Mechanisms for establishing second-order conditioning within the cerebellum are unknown. A TD implementation based on LTD assumes that PF inputs arising from second-order serial components express LTD even though the US is withheld. Hartell (1996) provides evidence favorable for such a mechanism.

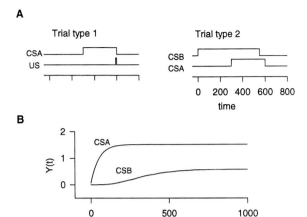

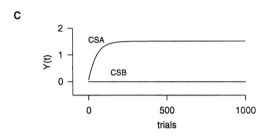

Figure 8: Simulated first- and second-order conditioning after Kehoe et al. (1981). (*A*) The training protocol. Trial-type 1: CSA is 300 msec in duration, and the CS–US interval is 290 msec. Trial type 2: CSB is 550 msec in duration, and the CSB–CSA interval is 300 msec. The two trial types are randomly mixed, and each occurs 500 times. (*B*) Simulated peak CR amplitude [$Y(t)$] to CSA and CSB as a function of trials. Note that $Y(t)$ to CSB increases with training, reflecting second-order conditioning. (*C*) With $\gamma = 0$ (equivalent to inactivation of the RN), second-order conditioning does not occur.

established, reinforced, and maintained among PCs that determine the timing and topography of CRs. The implementation incorporates recent anatomical findings, reviewed by Rosenfield and Moore (1995), that allow these PCs to receive the two components of the TD model's reinforcement operator, one component donated by the US and another component donated by $\dot{Y}(t) = Y(t) - Y(t - \Delta t)$ as feedback efference. The implementation lays the foundation for network simulations at the cellular level.

Acknowledgments

We thank the following individuals for helpful comments on various aspects of this article: Eli Brandt, Peter Dayan, Bernard Schreurs, and an anonymous reviewer.

References

Berthier, N.E. and J.W. Moore. 1986. Cerebellar Purkinje cell activity related to the classically conditioned nictitating membrane response. *Exp. Brain Res.* **63:** 341–350.

———. 1990. Activity of deep cerebellar nuclear cells during classical conditioning of nictitating membrane extension in rabbits. *Exp. Brain Res.* **83:** 44–54.

Berthier, N.E., A.G. Barto, and J.W. Moore. 1991. Linear systems analysis of the relationship between firing of deep cerebellar neurons and the classically conditioned nictitating membrane response in rabbits. *Biol. Cybern.* **65:** 99–105.

Blazis, D.E.J. and J.W. Moore. 1991. Conditioned inhibition of the nictitating membrane response in rabbits following hypothalamic and mesencephalic lesions. *Behav. Brain Res.* **46:** 71–81.

Bullock, D, J.C. Fiala, and S. Grossberg. 1994. A neural model of timed response learning in the cerebellum. *Neural Networks* **7:** 1101–1114.

Chen, C. and R.F. Thompson. 1995. Temporal specificity of long-term depression in parallel fiber-Purkinje synapses in rat cerebellar slice. *Learn. & Mem.* **2:** 185–198.

Clark, R.E. and D.G. Lavond. 1993. Reversible lesions of the red nucleus during acquisition and retention of a classically conditioned behavior in rabbits. *Behav. Neurosci.* **107:** 264–270.

Desmond, J.E. 1990. Temporally adaptive responses in neural models: The stimulus trace. In *Learning and computational neuroscience: Foundations of adaptive networks* (ed. M. Gabriel and J. Moore), pp. 421–461. MIT Press, Cambridge, MA.

Desmond, J.E. and J.W. Moore. 1988. Adaptive timing in neural networks: The conditioned response. *Biol. Cybern.* **58:** 405–415.

———. 1991a. Activity of red nucleus neurons during the classically conditioned rabbit nictitating membrane response. *Neurosci. Res.* **10:** 260–279.

———. 1991b. Altering the synchrony of stimulus trace processes: Tests of a neural-network model. *Biol. Cybern.* **65:** 161–169.

Edgley, S.A. and M. Lidierth. 1987. Discharges of cerebellar Golgi cells during locomotion in cats. *J. Physiol. (Lond.)* **392:** 315–332.

Eilers, J., G.J. Augustine, and A. Konnerth. 1995. Subthreshold synaptic Ca^{2+} signaling in fine dendrites and spines of cerebellar Purkinje neurons. *Nature* **373:** 155–158.

Fiala, J.C., S. Grossberg, and D. Bullock. 1996. Metabotropic glutamate receptor activation in cerebellar Purkinje cells as substrate for adaptive timing of the classically conditioned eye-blink response. *J. Neurosci.* **16:** 3760–3774.

Ghosh, A. and M.E. Greenberg. 1995. Calcium signaling in neurons: Molecular mechanisms and cellular consequences. *Science* **268:** 239–247.

Grossberg, S. and J.W.L. Merrill. 1996. The hippocampus and cerebellum in adaptively timed learning, recognition, and movement. *J. Cognit. Neurosci.* **8:** 257–277.

Gruart, A. and C.H. Yeo. 1995. Cerebellar cortex and eyeblink conditioning: Bilateral regulation of conditioned responses. *Exp. Brain Res.* **104:** 431–448.

Hardiman, M.J., N. Ramnani, and C.H. Yeo. 1996. Reversible inactivations of the cerebellum with muscimol prevent the acquisition and extinction of conditioned nictitating membrane responses in the rabbit. *Exp. Brain Res.* **110:** 235–247.

Hartell, N.A. 1996. Strong activation of parallel fibers produces localized calcium transients and a form of LTD that spreads to distant synapses. *Neuron* **16:** 601–610.

Hesslow, G. 1994. Inhibition of classically conditioned eyeblink responses by stimulation of the cerebellar cortex in the cat. *J. Physiol.* **476:** 245–256.

Ito, M. 1984. *The cerebellum and neural control*. Raven Press, New York, NY.

Kano, M., U. Rexhausen, J. Dreessen, and A. Konnerth. 1992. Synaptic excitation produces a long-lasting rebound potentiation of inhibitory synaptic signals in cerebellar Purkinje cells. *Nature* **356:** 601–604.

Kehoe, E.J., A.M. Feyer, and J.L. Moses. 1981. Second-order conditioning of the rabbit's nictitating membrane response as a function of the CS2–CS1 and CS1–US intervals. *Anim. Learn. & Behav.* **9:** 304–315.

Kim, J.J. and R.F. Thompson. 1997. Cerebellar circuits and synaptic mechanisms involved in classical eyeblink conditioning. *Trends Neurosci.* **20:** 177–181.

Klopf, A.H. 1988. A neuronal model of classical conditioning. *Psychobiology* **16:** 85–125.

Konnerth, A., J. Dreessen, and G.T. Augustine. 1992. Brief dendritic signals initiate long-lasting synaptic depression in cerebellar Purkinje cells. *Proc. Natl. Acad. Sci.* **89:** 7051–7055.

Millenson, J.R., E.J. Kehoe, and I. Gormezano. 1977. Classical conditioning of the rabbit's nictitating membrane response under fixed and mixed CS–US intervals. *Learn. & Motiv.* **8:** 351–366.

Moore, J.W. 1992. A mechanism for timing conditioned responses. In *Time, action, and cognition* (ed. F. Macar, V. Pouthas, and W.J. Friedman), pp. 229–238. Kluywer Academic Publishers, Dordrecht, The Netherlands.

Moore, J.W. and J.E. Desmond. 1992. A cerebellar neural network implementation of a temporally adaptive conditioned response. In *Learning and memory: The behavioral and biological substrates* (ed. I. Gormezano and E.A. Wasserman) pp. 347–368. Lawrence Erlbaum Associates, Hillsdale, NJ.

Moore, J.W., J.E. Desmond, N.E. Berthier, D.E.J. Blazis, R.S. Sutton, and A.G. Barto. 1986. Simulation of the classically conditioned nictitating membrane response by a neuron-like adaptive element: Response topography, neuronal firing, and interstimulus intervals. *Behav. Brain Res.* **21:** 143–154.

Moore, J.W., J.E. Desmond, and N.E. Berthier. 1989. Adaptively timed conditioned responses and the cerebellum: A neural network approach. *Biol. Cybern.* **62:** 17–28.

Perrett, S.P. and M.D. Mauk. 1995. Extinction of conditioned eyelid responses requires the anterior lobe of cerebellar cortex. *J. Neurosci.* **15:** 2074–2080.

Perrett, S.P., B.P. Ruiz, and M.D. Mauk. 1993. Cerebellar cortex lesions disrupt learning-dependent timing of conditioned eyelid responses. *J. Neurosci.* **13:** 1708–1718.

Ramnani, N. and C.H. Yeo. 1996. Reversible inactivations of the cerebellum prevent the extinction of conditioned nictitating membrane responses in rabbits. *J. Physiol.* **495:** 159–168.

Raymond, J.L., S.G. Lisberger, and M.D. Mauk. 1996. The cerebellum: A neuronal learning machine? *Science* **272:** 1126–1131.

Rescorla, R.A. and A.R. Wagner. 1972. A theory of Pavlovian conditioning: Variations in the effectiveness of reinforcement and nonreinforcement. In *Classical conditioning II: Current theory and research* (ed. A.H. Black and W.F. Prokasy), pp. 64–99. Appleton-Century-Crofts, New York, NY.

Richards, W.G., T.N. Ricciardi, and J.W. Moore. 1991. Activity of spinal trigeminal pars oralis and adjacent reticular formation units during differential conditioning of the rabbit nictitating membrane response. *Behav. Brain Res.* **44:** 195–204.

Rosenfield, M.E. and J.W. Moore. 1995. Connections to cerebellar cortex (Larsell's HVI) in the rabbit: A WGA-HRP study with implications for classical eyeblink conditioning. *Behav. Neurosci.* **109:** 1106–1118.

Schreurs, B.G. and D.L. Alkon. 1993. Rabbit cerebellar slice analysis of long-term depression and its role in classical conditioning. *Brain Res.* **631:** 235–240.

Schreurs, B.G., M.M. Oh, and D.L. Alkon. 1996. Pairing-specific long-term depression of Purkinje cell excitatory postsynaptic potentials results from a classical conditioning procedure in the rabbit cerebellar slice. *J. Neurophysiol.* **75:** 1051–1060.

Schreurs, B.G., D. Tomsic, P.A. Gusev, and D.L. Alkon. 1997. Dendritic excitability microzones and occluded long-term depression after classical conditioning of the rabbit's nictitating membrane response. *J. Neurophysiol.* **77:** 86–92.

Schultz, W., P. Dayan, and P.R. Montague. 1997. A neural substrate of prediction and reward. *Science* **275:** 1593–1599.

Sears, L.L. and J.E. Steinmetz. 1991. Dorsal accessory inferior olive activity diminishes during acquisition of the rabbit classically conditioned eyelid response. *Brain Res.* **545:** 114–122.

Sutton, R.S. and A.G. Barto. 1981. Toward a modern theory of adaptive networks: Expectation and prediction. *Psychol. Rev.* **88:** 135–171.

———. 1990. Time-derivative models of Pavlovian reinforcement. In *Learning and computational neuroscience: Foundations of adaptive networks* (ed. M. Gabriel and J. Moore), pp. 497–537. MIT Press, Cambridge, MA.

Thompson, R.F. 1986. The neurobiology of learning and memory. *Science* **233:** 941–947.

Thompson, R.F. and D.J. Krupa. 1994. Organization of memory traces in the mammalian brain. *Annu. Rev. Neurosci.* **17:** 519–549.

van Kan, P.L.E., A.R. Gibson, and J.C. Houk. 1993. Movement-related inputs to intermediate cerebellum of the monkey. *J. Neurophysiol.* **69:** 74–94.

Wells, G.R., M.J. Hardiman, and C.H. Yeo. 1989. Visual projections to the pontine nuclei of the rabbit: Orthograde and retrograde tracing studies with WGA-HRP. *J. Comp. Neurol.* **279:** 629–652.

Received February 18, 1997; revised version accepted April 30, 1997.

A Model of Pavlovian Eyelid Conditioning Based on the Synaptic Organization of the Cerebellum

Michael D. Mauk[1,3] and Nelson H. Donegan[2]

[1]Department of Neurobiology and Anatomy
The University of Texas Medical School at Houston
Houston, Texas 77030
[2]Department of Psychology
Yale University
New Haven, Connecticut 06520

Abstract

We present a model based on the synaptic and cellular organization of the cerebellum to derive a diverse range of phenomena observed in Pavlovian eyelid conditioning. These phenomena are addressed in terms of critical pathways and network properties, as well as the sites and rules for synaptic plasticity. The theory is based on four primary hypotheses: (1) Two cerebellar sites of plasticity are involved in conditioning: (a) bidirectional long-term depression/potentiation at granule cell synapses onto Purkinje cells ($gr \rightarrow Pkj$) in the cerebellar cortex and (b) bidirectional plasticity in the interpositus nucleus that is controlled by inhibitory inputs from Purkinje cells; (2) climbing fiber activity is regulated to an equilibrium level at which the net strength of $gr \rightarrow Pkj$ synapses remains constant unless an unexpected unconditioned stimulus (US) is presented or an expected US is omitted; (3) a time-varying representation of the conditioned stimulus (CS) in the cerebellar cortex permits the temporal discrimination required for conditioned response timing; and (4) the ability of a particular segment of the CS to be represented consistently across trials varies as a function of time since CS onset. This variation in across-trials consistency is thought to contribute to the ISI function. The model suggests several empirically testable predictions, some of which have been tested recently.

An initial challenge in understanding the neural mechanisms of associative learning phenomena is to identify the brain structures and pathways that are responsible for the acquisition and expression of a learned behavior. This difficult task generally involves a reductionistic strategy in which lesions, electrical stimulation, neural recordings, and neuroanatomical tracing techniques are used to identify neural structures and connections responsible for learning. A subsequent challenge is to identify the ways in which neural components interact to produce the range of observed learning phenomena. This requires an integrative approach in which critical features of the system are described with sufficient accuracy and detail to derive the target behaviors. One approach to this task is the use of hypothetical models that attempt to derive accounts for behavioral phenomena by combining features of the known synaptic organization and physiology of the brain regions thought to be involved with explicitly stated assumptions regarding how those regions are engaged by training. Such models can be useful, even before localization issues are completely resolved, by highlighting structural and functional properties that could produce learning, by generating testable predictions, and by pointing to the kinds of experiments that would be especially important for further progress.

Here, we outline a cerebellar model of Pavlovian eyelid conditioning based on the seminal cerebellar theories of Marr (1969) and Albus (1971). Our focus on the cerebellum and related brainstem structures is motivated by the diverse range of experiments suggesting that the cerebellum is

[3]Corresponding author.

an element in the pathway that is activated by the conditioned stimulus (CS) and that drives the expression of conditioned responses (CRs) (the CS → CR pathway). As outlined below, numerous studies suggest that the CS → CR pathway involves the particular pontine nuclei neurons activated by the CS, their mossy fiber projections to the cerebellum, the cerebellar projections to red nucleus, and the relevant motor nuclei (see Fig. 2, below). Other studies indicate that the unconditioned stimulus (US) activates climbing fiber inputs to the cerebellum. These observations have been interpreted as evidence that with appropriately timed mossy fiber and climbing fiber activation by the CS and US, respectively, the CS acquires the ability to activate the cerebellar interpositus nucleus neurons responsible for eliciting eyelid closure.

The primary support for these assumptions are (1) The retention of CRs is spared by decerebration (Mauk and Thompson 1987); (2) stimulation of appropriate regions of the anterior interpositus nucleus in trained *or untrained* animals elicits robust eyelid responses at modest stimulation parameters (McCormick and Thompson 1984); (3) lesions of the anterior interpositus nucleus abolish the expression of CRs (e.g., McCormick et al. 1982; McCormick and Thompson 1984; Yeo et al. 1985a; Welsh and Harvey 1989, 1992); (4) lesions of climbing fibers produce effects that mimic the absence of the US (McCormick et al. 1985; although, see Yeo et al. 1986); (5) lesions of mossy fibers can mimic removal of the CS (Steinmetz et al. 1987, 1988); (6) stimulation of mossy fibers and climbing fibers can substitute for the CS and US, respectively, to promote acquisition of CRs that are abolished by subsequent lesions of the anterior interpositus (Mauk et al. 1986; Steinmetz et al. 1989; Steinmentz 1990); and (7) reversible inactivation of the anterior interpositus nucleus retards learning (Krupa et al. 1993; Nordholm et al. 1993).

Thus, the view that the cerebellum is a necessary element of the CS → CR pathway is supported by evidence that information about the CS and US is conveyed to the cerebellum by separate afferent systems, that disrupting cerebellar output abolishes CRs, and that cerebellar output is capable of driving the expression of responses. Moreover, the convergence of CS and US information at the cerebellum combined with the ability of cerebellar output to elicit the expression of eyelid responses makes the cerebellum a strong candidate for the sites of plasticity that are responsible for the acquisition and expression of conditioned eyelid responses. A natural extension of this view is that certain behavioral properties of eyelid conditioning reflect the input/output processing properties of the cerebellum.

However, recent studies have challenged this cerebellar hypothesis and motivated the counterhypothesis that the cerebellum projects to, but is not a part of, a CS → CR pathway in the brain stem (Welsh and Harvey 1989, 1991; Kelly et al. 1990; Llinás and Welsh 1993; see also Bower and Kassel 1990). The influence that the cerebellum may have on these pathways is hypothesized to be modulatory, affecting either the performance of the responses (Welsh and Harvey 1991) or sensory processing (Gao et al. 1996). In this view, lesions of the cerebellum are thought to abolish CRs by disrupting the normal functioning of the critical brainstem pathways or of critical sensory discrimination (but, for other views, see Lavond et al. 1993 or Thompson and Krupa 1994).

Thus, whereas even the basic issue of cerebellar involvement in eyelid conditioning is not resolved to everyone's satisfaction, we suggest that another useful approach is to explore the feasibility of the mechanisms associated with each hypothesis. As a complement to additional empirical experiments evaluating whether the cerebellum (or brain stem) is responsible for eyelid conditioning, we can also ask whether the synaptic organization and physiology of the cerebellum (or brain stem) comprise a system with the capacity to generate the corpus of experimental data. To the extent that these questions yield awkward, biologically implausible answers, the hypothesis may be flawed. On the other hand, showing that the proposed cerebellar (or brain stem) mechanisms could generate the behavior in a biologically plausible manner would demonstrate that the system is, in principle, capable of explaining the facts of conditioning.

The development of the present model was driven by two well-characterized data sets: the rich literature describing in detail the synaptic organization of the cerebellum (e.g., Eccles et al. 1967; Ito 1984) and the extensively documented behavioral properties of eyelid conditioning (e.g., Wagner 1969; Gormezano 1972; Moore 1972; Siegel 1972; Gormezano et al. 1983, 1987; Kehoe and Napier 1991a,b; Napier et al. 1992; Kehoe et al. 1993a,b). Although we have incorporated known aspects of cerebellar anatomy and physiology into this model and have attempted to make as few gratuitous assumptions as possible, we emphasize the hypothetical nature of this work. Our goal is to examine the degree to which biologically reasonable accounts of conditioning can be obtained from known principles of cerebellar neural processing.

Basic Properties of Eyelid Conditioning

Acquisition of conditioned eyelid responses requires the paired presentation of the CS and US. Typically, a relatively neutral stimulus such as a tone serves as the CS, and a periorbital shock or an air puff directed at the eye serves as the US (Fig. 1A). Initially, the US elicits an eyelid response, whereas there is no measurable eyelid response to the CS. With repeated CS + US pairings, the CS acquires the ability to elicit conditioned eyelid responses (Fig. 1B,C) that are timed to peak near US onset (Fig. 1E). The rate of acquisition and the timing of the CRs are influenced strongly by the interstimulus interval (ISI) between CS and US onsets (Schneiderman and Gormezano 1964; Schneiderman 1966; Frey and Ross 1968; Smith 1968; Salafia et al. 1979). The CS onset must precede the US by at least 80 msec (i.e., ISI > 80 msec) but not by >2-3 sec for response to develop (Fig. 1D).

Core Assumptions and Basic Architecture of the Model

A basic organizing principle of the model is the hypothesis that the CS activates mossy fibers, the US activates climbing fibers, and output from the interpositus nucleus activates neurons that generate the conditioned eyelid responses (Fig. 2). In terms of the CS representation conveyed to the cerebellum, we assume that (1) each CS is encoded by activity in a subset of mossy fibers; (2) these subsets share common elements (active mossy fibers) to the extent the CSs are similar; and (3) the mossy fiber CS representation conveys minimal temporal information, no more than might be provided by three categories of mossy fiber inputs, those that show either tonic, phasic, or phasic/tonic responses to the CS.[1] The reinforcing properties of the US are assumed to be generated by the activation of climbing fibers that are specific for the US; for example, air puff to the eye activates climbing fibers from the facial regions of the inferior olive (IO) (Mauk et al. 1986). Finally, we assume that CS + US pairings can induce plasticity in the cerebellum and that the expression of CRs is mediated by increased activity in the appropriate sets of cells in the anterior interpositus nucleus of the cerebellum. Thus, the muscle groups involved in the CR are determined by the anterior interpositus nucleus neurons that are activated, and the magnitude of the response is determined by their degree of activity.

SYNAPTIC ORGANIZATION OF THE CEREBELLUM

The architecture of the cerebellum (Fig. 2)—the cell types, their connectivity, and their synaptic physiology—is relatively well known (Eccles et al. 1967; Palay 1974; Chan-Palay 1977; Ito 1984). As noted above, inputs to the cerebellum are conveyed by two classes of afferents: the mossy fibers and climbing fibers. Mossy fibers originate from numerous sources throughout the brain, and different mossy fibers are activated by virtually every sensory modality, including a large number that receive input from the cerebral cortex. Mossy fibers branch profusely in the cerebellar cortex, make excitatory synapses with Golgi cells and a large number of granule cells, and also send excitatory collaterals to the cerebellar nucleus cells. Granule cell axons (parallel fibers) course through the folium making excitatory contacts with numerous Purkinje cells, which in turn inhibit the cells of the cerebellar nuclei. Thus, the mossy fibers can influence the output cells in the cerebellar nuclei both by direct excitatory projections and by the granule/Purkinje pathway in the cerebellar cortex. Climbing fibers arise from the IO nuclei and project directly to cerebellar Purkinje cells. In contrast to the extensive divergence of mossy fibers, each climbing fiber contacts ~10 Purkinje cells and each Purkinje cell receives input from only one climbing fiber.

---

[1]Recordings from mossy fibers activated by auditory stimuli have demonstrated a limited amount of temporal variation during presentation of tones (Aitkin and Boyd 1978). In general, these neurons respond phasically at tone onset and offset, tonically, or phasic/tonic. These variations over time are not in a range that can directly account for the range CR timing supported by various ISIs.

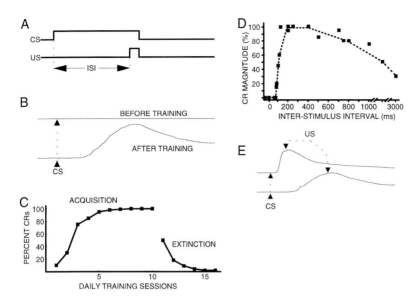

Figure 1: Behavioral properties of Pavlovian eyelid conditioning. (*A*) A schematic representation of a CS + US trial for a commonly used protocol (delay conditioning). The onset of the CS, a tone or a light, for example, precedes the onset of an US such as an air puff directed at the eye. The time between the onsets of the stimuli is the ISI. (*B*) Sample eyelid responses elicited by the presentation of a tone CS (onset depicted by the arrowheads). Before training there is no measurable eyelid response to the CS. After training the eyelid closes (upward deflection of the trace) in response to the CS, and the response peaks near the normal time of US onset (not shown). (*C*) Sample acquisition and extinction curves showing the increase in response probability during 10 days of CS + US training and the decline in response magnitude during 6 days of CS-alone extinction training. (*D*) The effects of the ISI on conditioning (the ISI function) taken from a number of published results (Schneiderman and Gormezano 1964; Schneiderman 1966; Smith 1968; Salafia et al. 1979). Each point represents the magnitude of conditioning for that ISI expressed as a percentage of the maximum responding observed in the particular paper. A negative ISI indicates the US precedes the CS, and a positive ISI indicates CS precedes the US. (*E*) Two sample responses from animals trained with a relatively short (*top*) and relatively longer (*bottom*) ISI. CS onsets are indicated by upward pointing arrowheads, and the US onsets by downward arrowheads. The latencies to onset and rise-times of the responses vary systematically with the ISI such that the responses peak near US onset.

THE SEMINAL CEREBELLAR THEORIES OF MARR AND ALBUS

On the basis of this synaptic organization, Marr (1969) and, later, Albus (1971) formulated similar theories of cerebellar function. Marr suggested that the mossy fibers convey to the cerebellum information about the context in which movements occur. The representation of this context was said to be recoded through the divergent mossy fiber projections onto the very large population of granule cells. Marr suggested that movements in that context could be reinforced if climbing fiber inputs increase the strength of coactive $gr \rightarrow Pkj$ synapses. Albus modified this theory by assuming that climbing fibers decrease the strength of coactive $gr \rightarrow Pkj$ synapses. Many studies have since provided support for the basic architecture of the Marr/Albus theories and, in particular, for the plasticity rule proposed by Albus. For example, coactivation of climbing fiber and granule cell inputs to the Purkinje cells can decrease the strength of the $gr \rightarrow Pkj$ synapses, a phenomenon called long-term depression (LTD) (Ito and Kano 1982; Ito et al. 1982; Ito 1988; Linden and Connor 1991; Linden et al. 1991; Schreurs and Alkon 1993, 1996).

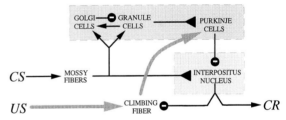

Figure 2: A schematic (and partial) representation of the synaptic organization of the cerebellum and its relation to the stimulus and response pathways involved in eyelid conditioning. The two classes of cerebellar afferents—mossy fibers and climbing fibers—are activated by the CS and US, respectively. Mossy fibers project directly to the output cells of the cerebellar nuclei and also to the granule and Golgi cells of the cerebellar cortex. The granule cells make excitatory synapses onto Purkinje cells that, in turn, inhibit the nucleus output cells. Granule cell activity in response to a particular mossy fiber input is influenced by interactions between the populations of granule and Golgi cells. In contrast, climbing fibers make spatially distributed connections onto a small number of Purkinje cells. The output cells of the cerebellar interpositus nucleus project to efferent pathways that are necessary for the expression of the conditioned eyelid responses. This output pathway also provides feedback inhibition of the climbing fibers.

An important extension of this rule allows for $gr \rightarrow Pkj$ synapses to increase in strength [i.e., to undergo long-term potentiation]. This assumption is based on experiments showing that $gr \rightarrow Pkj$ synapses are potentiated if they are active in the absence of climbing fiber inputs (Sakurai 1987; Shibuki and Okada 1992; Schreurs and Alkon 1993; Salin et al. 1996). In addition, lesion studies examining eyelid conditioning and adaptation of vestibular reflexes (Robinson 1976; McCormick and Thompson 1984), as well as unit recordings during adaptation of wrist movements (Thach 1980; Thach et al. 1992), have produced results consistent with the spirit of the Marr and Albus theories. The present model is an elaboration of these theories.

Hypothesis 1: Conditioning Involves Two Cerebellar Sites of Synaptic Plasticity

We hypothesize that two sites of synaptic plasticity in the cerebellum are responsible for the acquisition and expression of conditioned eyelid responses: (1) the $gr \rightarrow Pkj$ synapses in the cerebellar cortex and (2) the mossy fiber synapses in the interpositus nucleus ($mf \rightarrow nuc$). Motor learning-related plasticity at one or both of these sites has been proposed previously (Miles and Lisberger 1981; Ito 1982; Thompson 1986; Lisberger 1988; Perrett et al. 1993; Perrett and Mauk 1995; Raymond et al. 1996; Mauk 1997). The novel aspects of our hypothesis are found in the explicitly stated rules for plasticity at both sites.

1.1. BIDIRECTIONAL PLASTICITY AT $gr \rightarrow Pkj$ SYNAPSES IN THE CEREBELLAR CORTEX

From the assumptions of Albus (1971) and subsequent empirical findings (Sakurai 1987; Ito 1988; Kano and Kato 1988; Linden and Connor 1991; Linden et al. 1991; Schreurs and Alkon 1993, 1996), we adopt the following bidirectional LTD/LTP rule: $gr \rightarrow Pkj$ synapses decrease in strength when active in the presence of a climbing fiber input (LTD) and increase in strength when active in the absence of a climbing fiber input (LTP). Although the exact timing required of the two inputs has not been precisely identified (Kano and Kato 1988; Chen and Thompson 1995; Schreurs and Alkon 1996), the model requires only that LTD occur when the inputs are active at approximately the same time.

1.2. BIDIRECTIONAL PLASTICITY AT $mf \rightarrow nuc$ SYNAPSES CONTROLLED BY PURKINJE CELL INPUTS

At $mf \rightarrow nuc$ synapses we assume a bidirectional plasticity that is dependent on the inhibitory Purkinje cell input: Synapses increase in strength (LTP_{nuc}) when active during *transient* decreases in Purkinje activity; that is, this rule is not engaged by the mere absence of Purkinje input. Conversely, we assume that $mf \rightarrow nuc$ synapses decrease in strength (LTD_{nuc}) when active in the presence of strong inhibitory Purkinje cell input. We know of only one study in which plasticity in the nucleus was tested—high-frequency stimulation of mossy fibers enhanced a monosynaptic field potential recorded in cerebellar nuclei in vivo (Racine et al. 1986). Other, indirect evidence suggests that bidirectional changes in these synapses are important for adaptation of the vestibulo-ocular reflex (e.g., Miles and Lisberger 1981; Lisberger 1988) and for eyelid conditioning (Thompson 1986; Perrett et al. 1993; Perrett and Mauk 1995; Raymond et al. 1996; Mauk 1997). Thus, although experimental findings indirectly support the idea that $mf \rightarrow nuc$ synapses are modifiable, the Purkinje-dependent rule for plasticity at $mf \rightarrow nuc$ synapses is an assumption whose consequences we evaluate in the model.

1.3. FACTORS INFLUENCING THE INDUCTION OF PLASTICITY

The findings of Llinás and Mühlethaler (1988) suggest a concrete but speculative mechanism for inducing LTP_{nuc} and LTD_{nuc} at $mf \rightarrow nuc$ synapses that is consistent with our assumptions. Recordings from cerebellar nucleus cells revealed a calcium conductance whose activation required depolarization from a hyperpolarized state, rather than depolarization from the resting potential. Given that Purkinje cells are normally active at high rates, it seems possible that this calcium conductance could be activated by *transient* decreases in the ongoing inhibitory input from Purkinje cells and that the resulting increase in intracellular calcium could induce LTP_{nuc} in coactive $mf \rightarrow nuc$ synapses. Drawing parallels with LTP and LTD in the hippocampus and neocortex (Artola et al.

1990; Mulkey and Malenka 1992; Dudek and Bear 1993), mf → nuc synapses may increase in strength when coactive during high levels of calcium and decrease in strength when active during lower levels of calcium, as may occur during strong inhibitory input from Purkinje cells (see Shibuki and Okada 1992).

Recordings have shown that Purkinje cells are normally active at high rates, and the corresponding tonic inhibition of nucleus cells highlights an important implication of our plasticity assumptions. Whereas LTD/LTP at the $gr \to Pkj$ synapses depends on the patterns of stimuli conveyed by mossy fiber and climbing fiber inputs, LTP_{nuc}/LTD_{nuc} at the mf → nuc synapses is controlled by mossy fiber collateral inputs and by inhibitory input from Purkinje cells. Thus, lesions of the cerebellar cortex should prevent further plasticity at mf → nuc synapses. Without inhibition from Purkinje cells, our assumptions suggest that LTD_{nuc} would not be possible. Similarly, without Purkinje cells there cannot be a transient release from inhibition that our assumptions require for the induction of LTP_{nuc}.

1.4. A TRIALS-LEVEL DESCRIPTION OF ACQUISITION, EXPRESSION, AND EXTINCTION OF CRs

The above assumptions suggest the following trials-level account for cerebellar involvement in eyelid conditioning (Fig. 3). Prior to training, the mossy fibers activated by the CS only weakly excite the nucleus cells. This excitation is counteracted by the ongoing inhibitory input from Purkinje cells, which is not changed by the CS. Consequently, the initial acquisition of CRs results from the induction of LTD at $gr \to Pkj$ synapses that are activated by the CS during a US-evoked climbing fiber input to the Purkinje cell. With repeated CS + US pairings, the LTD at these $gr \to Pkj$ synapses would decrease the average strength of the $gr \to Pkj$ synapses active during the CS. Because

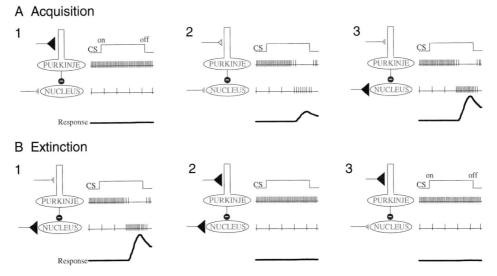

Figure 3: A schematic representation of the events hypothesized to be associated with the acquisition and extinction of CRs. Each of the six panels shows the granule cell synapses onto Purkinje cells ($gr \to Pkj$) and the mossy fiber synapses onto the nucleus cells (mf → nuc), where the relative strength of the synapse is indicated by the size and shading (stronger synapses are larger and darker). The *right* half of each panel shows the assumed activity of the Purkinje and nucleus cells during the presentation of a CS. The response that would be produced is shown schematically at *bottom* of each panel. (A) Acquisition: (*1*) presentation of a CS to an untrained animal; (*2*) early in training, LTD induced by paired presentation of CS + US has decreased the strength of the $gr \to Pkj$ synapses; (*3*) with additional training, the induction of LTP at the mf → nuc synapses combines with the learned pause in Purkinje activity to produce robust CRs. (B) Extinction: (*1*) At the beginning of extinction training, CRs are produced by plasticity in both the cerebellar cortex and cerebellar nucleus; (*2*) i.e., when the decline in CRs is nearly complete, the induction of LTP at $gr \to Pkj$ synapses has restored robust Purkinje cell activity during the CS, preventing the expression of CRs despite the relative strength of the mf → nuc synapses; (*3*) at this point, presentation of additional CS alone extinction trials would induce LTD at mf → nuc synapses.

Purkinje cell activity presumably reflects the net effects of excitatory input from granule cells and ongoing inhibitory input from stellate and basket cells, training should lead to an acquired decrease in Purkinje activity and a phasic disinhibition of the interpositus cells *during the CS* (Fig. 3A, 1 and 2). This would have the dual effects of increasing the likelihood of eliciting CRs and inducing LTP_{nuc} at CS-activated mf → nuc synapses (Fig. 3A, 3), which would further increase the ability of the CS to generate a conditioned eyelid response.

In contrast, extinction can be explained by the LTP (reversal of LTD) induced at *gr → Pkj* synapses during nonreinforcement of a trained CS, which results from CS activation of granule cells in the absence of climbing fiber input. This would eliminate the acquired pause in Purkinje activity, reinstate the inhibition of nucleus cells during the CS, diminish the CR, and promote the induction of LTD_{nuc} (reverse LTP_{nuc}) at the mf → nuc synapses (Fig. 3B, 1–3). These changes at *gr → Pkj* synapses in the cortex and mf → nuc synapses in the nucleus would each act to diminish the ability of the CS to elicit CRs.

Thus, for both acquisition and extinction, our assumptions predict that plasticity is first induced in the cortex at *gr → Pkj* synapses; this, in turn, allows subsequent training trials to induce changes in the opposite direction at mf → nuc synapses. Because Purkinje cells inhibit nucleus cells, decremental changes in cortical synapses and incremental changes in nucleus synapses work together to increase the magnitude of the CR, whereas incremental changes in the cortex and decremental changes in the nucleus lead to response extinction.

Hypothesis 2: Climbing Fiber Activity Is Self-Regulated to an Equilibrium Level

A tacit assumption common to most models of learning is that in the absence of the CS, US, or context cues associated with the US, there are no opportunities for reinforcement or nonreinforcement. Similarly, most neural networks permit plasticity only during training trials and not, for example, between trials or during probe trials. In contrast, realistic neural models of learning must confront the constraints that (1) the neurons whose activity represents the CS and the US can be active in the intervals between training trials or training sessions and (2) the rules for plasticity cannot be arbitrarily disengaged during these times. With the cerebellum, for example, climbing fibers display low and fairly regular rates of spontaneous activity (see Keating and Thach 1995). Moreover, it seems implausible that the mossy fiber and granule cells that encode the CS are always silent in the absence of the CS. This would impose the equally unrealistic requirement that every stimulus is encoded by a unique subset of mossy fibers and granule cells that are active if (and only if) the stimulus is present. These issues pose difficult questions. For example, what prevents the fortuitous pairings of granule cell and climbing fiber activity from inducing LTD at *gr → Pkj* synapses to the point of saturation? How can 100 US-evoked climbing fiber inputs during a training session convey information against a background of 3600 spontaneous climbing fiber inputs during the same period and >80,000 before the next session? Moreover, if the granule cells that encode a CS are also occasionally active in the absence of the CS or US, what prevents spontaneous extinction of CRs between sessions?

This section describes a two-part hypothesis to address these issues. First, the combination of spontaneous mossy and climbing fiber activity suggests that plasticity at *gr → Pkj* synapses is best understood as a stochastic process in which net changes in strength depend on the probability that synapse activation is paired with a climbing fiber input. Second, we propose that the inhibitory pathways from cerebellar nuclei to climbing fiber cell bodies in the IO (Groenewegen and Voogd 1977; Groenewegen et al. 1979; Oscarsson 1980; Houk and Gibson 1986; Anderson et al. 1988; Angaut and Sotelo 1989; de Zeeuw et al. 1989; Ruigrok and Voogd 1990) provide a form of negative feedback that balances the likelihood of LTD and LTP events such that the net strength of *gr → Pkj* synapses remains relatively constant under all but two conditions: when a CS occurs with an unexpected US—as during acquisition—and when a CS signals the US, but the US is omitted—as during extinction.

2.1. PROBABILISTIC LTD/LTP AND EQUILIBRIUM CLIMBING FIBER ACTIVITY

Bidirectional LTD/LTP implies that *gr → Pkj* synapses will undergo plasticity whenever they are active ("active" might mean one spike or a burst of spikes); they undergo LTD when there are chance pairings of granule cell and climbing fiber activity and undergo LTP when granule cell activity occurs alone. Thus, the *rate* of net change of a $gr \rightarrow Pkj$ synapse will depend on its activity, and the expected direction of change will depend on the *probability* that its activity is paired with a climbing fiber input. When the probability of pairing is high, the expected result is a net decrease in synaptic strength (LTD), whereas a net increase (LTP) is expected when the pairing probability is low (i.e., granule cell activity in the absence of a climbing fiber input). Between these two extremes there exists an intermediate probability where the effects of LTD and LTP balance and the expected net change in synaptic strength is zero. At this level of climbing fiber activity, any single instance of synapse activity will result in an increment or decrement in strength, but these changes occur in proportions for which the expected net change is zero. For convenience we refer to this equilibrium level of climbing fiber activity as CF^{Eq}.

The ability to maintain the spontaneous climbing fiber activity at CF^{Eq} could serve an important role in maintaining a relatively constant pattern of synaptic weights as established by previous motor learning experience. Outside of the conditioning context, any $gr \rightarrow Pkj$ synaptic activity would be equally likely to promote an increase or decrease in strength, with the expected net change being zero. However, even slight deviations from CF^{Eq}, if persistent, would cause the strength of all $gr \rightarrow Pkj$ synapses to drift and to eventually saturate at their minimum or maximum values.

To regulate climbing fiber activity with sufficient precision to achieve stability would appear to require coupling the average $gr \rightarrow Pkj$ synaptic weight, as indexed by activity of the Purkinje cells, to the modulation of climbing fiber activity by negative feedback. A central idea we advance here is that the inhibition of climbing fibers by cerebellar output (Fig. 2) represents this type of negative feedback and could regulate spontaneous climbing fiber activity to CF^{Eq}. To illustrate, consider the consequences of spontaneous climbing fiber activity above CF^{Eq} (Fig. 4A). Chance pairings would occur too frequently in that LTD would prevail over LTP, resulting in a net decrease in synaptic strengths. The decrease in Purkinje cell activity would increase the activity of cerebellar nucleus cells through reduced inhibition from Purkinje cells. The increased inhibitory output from the nucleus cells would, in turn, cause a corresponding decrease in the activity of climbing fibers.

Thus, the expected consequence of excess climbing fiber activity is an increased inhibition of climbing fibers resulting from a net decrease in the strength of the $gr \rightarrow Pkj$ synapses. Inappropriately low rates of spontaneous climbing fiber activity could be similarly counteracted by an inverse series of events. Consequently, climbing fiber activity could be driven to a level at which LTD and LTP events balance and the expected net change in $gr \rightarrow Pkj$ synapses is zero. A recent mathematical analysis supports these ideas and suggests that under these conditions LTP and LTD at the $mf \rightarrow nuc$ synapses also balance (G.T. Kenyon, J.F. Medina, and M.D. Mauk, in prep.).

2.2. THE CONSEQUENCES OF PROBABILISTC PAIRINGS OF MOSSY FIBER AND CLIMBING FIBER INPUTS

We can now elaborate on the previous trials-level description of acquisition, retention, and extinction of CRs by considering the consequences of presenting a CS under a variety of conditions. The left ordinate of Figure 4B shows the probability that activity of a $gr \rightarrow Pkj$ synapse is paired with a climbing fiber input, the right ordinate shows the corresponding change in synaptic strength under bidirectional LTD/LTP rules, and the abscissa represents time. The horizontal boundry between the light and dark gray in the graph depicts an arbitrarily chosen value for CF^{Eq}. In an untrained animal, because the $gr \rightarrow Pkj$ synapses activated by a CS-alone trial should, like all other $gr \rightarrow Pkj$ synapses, be reinforced at CF^{Eq} the expected net change in their synaptic weights should be zero (Fig. 4B, 1). When this same CS is paired with a US, the probability of pairing increases during the trial above that produced by CF^{Eq}, owing to the excitatory input to the climbing fibers added by the US. The expected results are a net decrement in strength of the $gr \rightarrow Pkj$ synapses activated by the CS and acquisition of CRs (Fig. 4B, 2, and C). As acquisition proceeds, the CS would exhibit an increased ability to elicit eyelid responses and to inhibit climbing fibers. This would diminish the ability of the US to activate climbing fibers and return the probability of pairing during the CS + US trial to that produced by CF^{Eq}. At this point the excitatory input to the climbing fibers from the US is exactly counteracted by CR-related inhibition of the climbing fibers.

Figure 4B (3) illustrates how the modulation of

climbing fibers could be important for extinction as well. During an unreinforced CS, climbing fiber activity should fall below CF^{Eq} owing to the CR-related inhibition of climbing fibers that is now unopposed by the excitatory drive provided by the US. Here, the expected change in synaptic strength is a net increase, which would increase Purkinje activity during the CS, restore inhibition of nucleus cells during the CS, and produce extinction of CRs (Fig. 4C). The CS-generated inhibition of climbing fibers would decrease as the CRs diminish over the course of extinction training, and the expected changes in $gr \rightarrow Pkj$ synaptic strength should again be zero when the probability of pairing returns to that produced by CF^{Eq}. Thus, the result of either acquisition or extinction is that a set of synaptic weights changes such that climbing fiber equilibrium is achieved in the presence of background cues alone and during the CS. Moreover, under our assumptions, the induction of plasticity at $mf \rightarrow nuc$ synapses requires the initial induction of plasticity at $gr \rightarrow Pkj$ synapses. Consequently, the above arguments for $gr \rightarrow Pkj$ synapses apply indirectly for $mf \rightarrow nuc$ synapses as well.

2.3. CS + US TRIALS PROMOTE TWO PHASES OF SYNAPTIC PLASTICITY

Here, we extend the trials-level description further by distinguishing between events that are initiated during the trial from those that occur afterwards. The need for this distinction arises from the prediction, implied by an equilibrium level of climbing fiber activity, that each CS + US trial initiates two phases of plasticity (Fig. 5). One phase involves the LTD induced during the trial at $gr \rightarrow Pkj$ synapses activated by the CS. Because this LTD produces a net decrease in the synaptic input to the Purkinje cell, climbing fiber activity will decrease below CF^{Eq}. Thus, during a period following the trial in which climbing fiber activity remains depressed, there will be a second phase of plasticity in which active $gr \rightarrow Pkj$ synapses undergo more LTP than LTD until climbing fiber equilibrium is restored. Each $gr \rightarrow Pkj$ synapse will undergo this net increase in strength proportional to its background activity following the trial. Thus, the expected net change in each $gr \rightarrow Pkj$ synapse after a CS + US trial and after the subsequent return

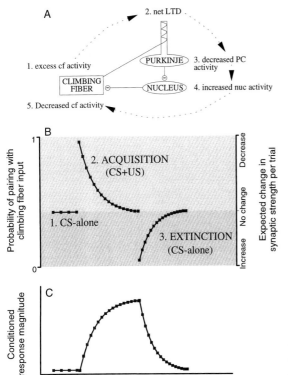

Figure 4: The self-regulating equilibrium of climbing fiber activity and its influence on acquisition and extinction. (*A*) The pathways involved in the feedback regulation of climbing fiber activity to the equilibrium level at which the expected net change in $gr \rightarrow Pkj$ synapses is zero. This example begins with a level of climbing fiber activity above equilibrium (*1*). This produces more LTD events than LTP, and the $gr \rightarrow Pkj$ synaptic weights gradually decrease (*2*). This decreases Purkinje cell activity (*3*), which decreases inhibition and increases the activity of nucleus cells (*4*). The resulting increase in inhibitory input brings climbing fiber activity back down to the equilibrium level. (*B*) The effect of self-regulating climbing fiber equilibrium on acquisition and extinction. The ordinate at *left* shows the probability than any given activation of a $gr \rightarrow Pkj$ synapse is paired with climbing fiber activity. The ordinate at *right* shows the corresponding net effect on the strength of $gr \rightarrow Pkj$ synapses. The horizontal line indicates the equilibrium level of climbing fiber activity at which LTD and LTP balance. Because climbing fiber activity is regulated to this level, the $gr \rightarrow Pkj$ synapses activated by a CS in an untrained level show no change (*1*). The presentation of a CS + US trial activates those $gr \rightarrow Pkj$ synapses when the probability of climbing fiber activity is much higher owing to the US (*2*). As acquisition proceeds, the response-related inhibition of climbing fibers counteracts the US-evoked excitation of climbing fibers, and the probability that the $gr \rightarrow Pkj$ synaptic activity associated with CS will be paired with a climbing fiber input decreases to the equilibrium level. During CS-alone extinction training, climbing fiber activity during the CS falls below the equilibrium level owing to the response-related inhibition that is unopposed by excitation from the US (*3*). As CRs decline, so does the response-related inhibition of climbing fibers, and thus the climbing fiber activity during the trial gradually returns to the equilibrium level.

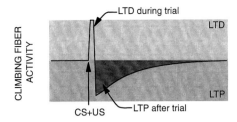

Figure 5: The two phases of plasticity induced by a CS + US trial that are suggested by a self-regulating equilibrium of climbing fiber activity. The ordinate depicts climbing fiber activity, and the abscissa depicts time. As in Fig. 4B, the horizontal line represents the equilibrium level of climbing fiber activity. Before the trial, spontaneous climbing fiber activity is in equilibrium. During the CS + US trial, the US increases climbing fiber activity above equilibrium and the $gr \rightarrow Pkj$ synapses activated by the CS undergo LTD. This decreases Purkinje (and thus climbing fiber) activity below equilibrium. Therefore, a second phase of plasticity occurs after the trial, during which $gr \rightarrow Pkj$ synapses tend to undergo LTP with a frequency proportional to their background activity.

to equilibrium should be proportional to the difference between its activity when the US-elicited climbing fiber input arrives during the CS (which produces LTD) and its background level of activity after the trial (which produces LTP). Synapses that are more active than normal when the US is presented undergo more LTD than LTP and display a net decrease in strength. The opposite holds for synapses that are less active than normal during the CS. It is important to note that this second phase of plasticity is attributable entirely to factors that maintain climbing fiber equilibrium and is not a *direct* product of post-trial US processing.

Hypothesis 3: A Time-Varying Granule Cell Representation of the CS Mediates CR Timing

In the next two sections we expand the trials-level description to include real-time, ISI-related phenomena. First, we address the ISI-dependent timing of CRs by hypothesizing a cerebellar mechanism that produces a time-varying representation of the CS. In the subsequent section we suggest how this mechanism may also provide an account for the relationship between rate of learning and the ISI (i.e., the ISI function).

3.1. A GRANULE CELL CS-REPRESENTATION THAT PROVIDES TEMPORAL DISCRIMINATION

The ISI not only determines whether CRs are acquired but also influences their timing (Schneiderman and Gormezano 1964; Schneiderman 1966; Frey and Ross 1968; Smith et al. 1969). Several studies have provided evidence that response timing cannot be mediated by a unitary, time-invariant CS representation. For example, Millenson et al. (1977) showed that when rabbits were trained using a single CS that was reinforced at two ISIs (e.g., the ISI was 200 msec for half the trials and was 700 msec for the remaining trials), they acquired CRs with two peaks, each peak corresponding to one of the training ISIs. Clearly, these data require an ability to discriminate different times during the CS, that is, that the CS representation consists of a time-varying series of discriminable features that can be differentially reinforced (Millenson et al. 1977; Mauk and Ruiz 1992; Perrett et al. 1993). We address these requirements by proposing that different subsets of granule cells become active at different times during the CS and that this provides the temporal discrimination necessary for CR timing.

In Figure 6B, a generic form of the proposed time-varying granule cell representation of a CS is contrasted with a time invariant representation (Fig. 6A). In both panels, the ordinate represents a hypothetical population of granule cells and the abscissa represents time during a CS. Because the time-invariant CS representation is similar at different times during the stimulus, it provides no basis for temporal coding. Training at any effective ISI would modify the same synapses. This would preclude the possibility of producing double peaked responses with training using two ISIs. Moreover, because these modified $gr \rightarrow Pkj$ synapses are active soon after CS onset, the CRs should show relatively short latencies to onset. These onset latencies could vary with the ISI only to the extent that associative strength in the nucleus can influence the rise time to the threshold for eliciting a response. In contrast, a time-varying representation as depicted in Figure 6B would encode not just the presence of the CS, but could encode different temporal segments of the CS. CS + US pairings with a particular ISI would modify a subset of $gr \rightarrow Pkj$ synapses whose activation occurs near US onset, because both the acquired decrease in Purkinje activity during the CS and the associated CRs would be delayed until just before US onset

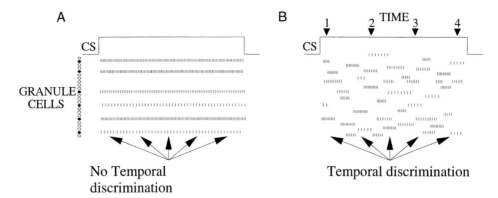

Figure 6: Time-invariant and time-varying granule cell activity during a CS. Both panels show hypothetical raster plots of the activity of a population of granule cells (ordinate) as a function of time during the presentation of a CS (abscissa). (*A*) A time-invariant representation of the CS in which a set subset of granule cells is activated throughout the extent of the stimulus. This representation provides no temporal discrimination. Training at any ISI would produce responses with the same short latency to onset and to peak. (*B*) A time-varying representation in which different subsets of granule cells become active at different times during the CS. Reinforcing this representation at a particular ISI would produce responses timed to peak near US onset.

when the greatest proportion of "conditioned" gr → Pkj synapses would be active. This type of CS representation could mediate the kinds of response timing that have been observed in the literature, such as double peaked responses with training using a single CS and two training ISIs.

3.2. A MECHANISM FOR TIME-VARYING ACTIVATION OF GRANULE CELLS

We also propose that the inhibitory influence of Golgi cells is a key factor in the production of a time-varying granule cell representation of CSs. If granule cells received input only from mossy fibers, then the subsets active during a CS could vary only to the extent that the mossy fiber activity varies during the CS (see Footnote 1). However, granule cells are also inhibited by Golgi cells, which receive excitatory input from both mossy fibers and granule cells (see Fig. 2). This synaptic arrangement suggests that the Golgi cells allow the *ongoing* activation of granule cells (by mossy fibers) to be modulated by *previous* activity of other mossy fibers and granule cells (see Fig. 6B). Thus, the granule cell activity at any time helps shape which granule cells will be activated next. For example, the granule cell activity at time "1" in Figure 6B would help determine the activity at time "2." Likewise, time "2" activity would influence subsequent activity at time "3," and so on. This would produce a characteristic variation in the granule cell activity throughout a stimulus.

Hypothesis 4: Across-Trials Consistency of the CS Representation Determines the ISI Function

In addressing the relationship between the ISI and rate of acquisition, most theories have adopted assumptions similar to those proposed by Hull (1943). (1) A CS activates a characteristic neural representation whose strength varies as a function of CS duration or trace interval, (2) the ability of the US to modify synapses will be proportional to the strength of the CS trace at the time of reinforcement, and, thus, (3) variation of the trace strength during the CS determines the ISI function. Here, we advance an hypothesis that is somewhat different than Hull's: The ability of granule cell activity to provide a consistent representation of the CS from one trial to the next (across-trials consistency) varies throughout the duration of the CS. We propose that the across-trials consistency of the activity representing the CS is an inverted U-shaped function of the time from CS onset.

4.1. THE NEED FOR ACROSS-TRIALS CONSISTENCY IN THE CS REPRESENTATION

Conditioning requires a degree of across-trials consistency in the CS representation because a CS can only elicit CRs if it activates, within a narrow window of time, a sufficient number of the synapses that were modified by previous CS + US pairings. Given our assumption that the CS is represented as a sequence of granule cell subsets active at different times during the stimulus, it is necessary that the ordering of this sequence be consistent from one trial to the next. Specifically, the time-varying pattern of granule cell activity, as schematized in Figure 6B, must be consistent from one trial to the next over the range of ISIs effective in producing conditioning. Consequently, decreases in across-trials consistency of granule activity should degrade the rate of conditioning, and variations in across-trials consistency for different segments of the CS would result in different rates of learning for different ISIs.

4.2. THE INFLUENCE OF NOISE ON THE TIME-VARYING CS REPRESENTATION

As a useful starting point, we assume that in the absence of the CS (i.e., during the inter-trial interval) there is a relative lack of consistency in the pattern of granule cell activity, despite the static contextual stimuli that may be represented by characteristic mossy fiber activity. This lack of consistency in the granule cell activity between trials would be responsible for the fact that US-alone presentations do not promote acquisition of conditioned eyelid responses to the context. Although the climbing fiber inputs activated by the US would induce LTD in $gr \rightarrow Pkj$ synapses that happen to be active at the time, substantially different subsets of synapses would be modified each trial. Moreover, the induction of LTD at these synapses would move climbing fiber activity below the equilibrium level, and all synapses would subsequently be expected to undergo LTP proportional to their background level of activity. The critical point is that without across-trials consistency each synapse's background activity determines both its likelihood of undergoing LTD (during the US) and undergoing LTP (as equilibrium is restored). Because each synapse would be as likely to decrease as increase in strength, no net synaptic changes should occur. In our view, the fact that the CS must precede the US by ~70–80 msec for conditioning to occur (Fig. 1D) may reflect the CS exposure required to promote sufficient levels of across-trials consistency in the granule cell activity. Similarly, optimal ISIs represent the amount of CS exposure required to produce the highest levels of consistency across trials (Fig. 7).

Why then does the ability of the CS to support conditioning decline as the ISI increases? This may

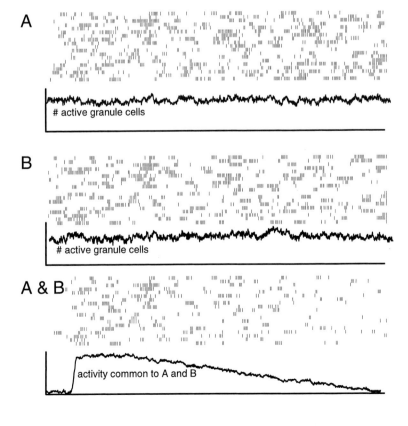

Figure 7: A schematic representation of across-trials consistency in the activity of granule cells activated by the CS. The two panels at *top* show hypothetical raster plots of granule cell activity as a function of time during the presentation of the same CS. For both panels the graph depicts a measure of the total amount of granule cell activity, which does not change appreciably during the CS. The third panel (*A&B*) shows the granule cell activity that is common to both trials, which is a measure of the across-trials consistency of granule activity. The graph associated with the panel at *bottom* shows that the common activity or across-trials consistency first increases then decreases during the CS. We propose that this across-trials consistency contributes to the conditionability of the CS at each time and therefore helps produce the ISI function.

reflect an inherent tendency for small amounts of inconsistency early in the CS to be amplified into large amounts of inconsistency in later portions of the CS (see Lorenz 1963). This tendency could arise from the way in which the activity of each cell at any given time is determined by previous activity in many other cells. A spurious spike in a granule cell may subsequently lead to spurious activity in Golgi cells that, given the divergence of their projections, could then lead to even more spurious activity (or lack of activity) in other granule cells. Thus, the degradation in consistency produced by a uniform noise source can accumulate and lead to a time-dependent decrease in the across-trials consistency of the neurons activated by a CS. The sources of noise could be as subtle as the statistical nature of transmitter release and ion channel openings/closings. In this view, long ISIs are ineffective for the same reasons that repeated US-alone trials do not support eyelid conditioning to the context; the US modifies the subset of $gr \to Pkj$ synapses that happens to be active, but these subsets are sufficiently different from trial to trial to prevent the development of a CR.

Accounts of Training and Physiological Manipulations on Behavioral Outcomes

We have proposed four interrelated hypotheses regarding cerebellar mechanisms of eyelid conditioning, which are summarized schematically in Figure 8. In this section we show how the assumptions interact to provide detailed accounts for a wide range of eyelid conditioning.

ACQUISITION, EXTINCTION, AND CLIMBING FIBER EQUILIBRIUM

Together, the four hypotheses imply that eyelid conditioning is mediated by interactions between (1) the rules for inducing synaptic plasticity at two cerebellar sites (cortex and nuclei), (2) the factors influencing the activity of climbing fibers, and (3) the nature of the CS representation in the cerebellum. Within the context of our assumptions, the cerebellum is a self-regulating system that maintains an equilibrium at which synapses are often modified, but the *net* change in strength of each synapse is zero. Pavlovian conditioning procedures can activate cerebellar afferents in ways that disturb this equilibrium and induce synaptic weight changes that mediate CRs.

Although theories of Pavlovian conditioning generally focus on events that occur during the CS + US trial, the present model suggests that it is also important to consider the equilibrium that exists before training and the events that occur after each trial. The activity of cerebellar inputs and cerebellar neurons elicited by training trials occurs against a background of ongoing mossy fiber and granule cell activity and an equilibrium level of climbing fiber activity (CF^{Eq}). Paired CS + US trials can provide conditions that disrupt this equilibrium and promote acquisition of CRs. These conditions are (1) repeated, US-evoked increases in climbing fiber activity above CF^{Eq}, during (2) CS-evoked granule cell activity that consistently differs from its background level. Extinction of CRs also occurs owing to a return to climbing fiber equilibrium. During the presentation of a CS-alone trial to a trained animal, the CR-related inhibition of climbing fibers, which is now unopposed by excitatory input from the US, brings the spontaneous climbing fiber activity below CF^{Eq}. This leads to net LTP at $gr \to Pkj$ synapses and the gradual decrease in the pause in Purkinje cell activity during the CS.

In the preceding sections, we have shown how the model can account for simple acquisition and extinction of CRs, as well as for the existence of spontaneous climbing fiber activity. We now explore additional and more detailed predictions of the model.

RELATIVE DISTRIBUTION OF PLASTICITY BETWEEN CORTEX AND NUCLEUS

Our assumptions imply that the relative distribution of plasticity can vary as a function of the amount and type of training (Mauk 1997). During acquisition, the paired presentation of CS and US would first induce LTD at the CS-activated $gr \to Pkj$ synapses. The resulting training-induced pause in Purkinje activity during the CS would contribute both to the expression of CRs and to the induction of LTP_{nuc} at CS-activated mf $\to$ nuc synapses. Our assumptions therefore permit, but do not demand, that CRs could appear before significant plasticity at mf $\to$ nuc synapses occurs. However, with continued training, responses would be mediated by plasticity in both the cortex and nucleus. This suggests that the degree to which CRs are retained following lesions of the cerebellar

Figure 8: A schematic summary of the four main hypotheses that comprise the model. We propose: (1*a*) Plasticity at *gr* → *Pkj* synapses is guided by climbing fiber input—synapses active during a climbing fiber input decrease in strength, and those active without a climbing fiber input increase in strength; (1*b*) plasticity at mf → nuc synapses is guided by inhibitory input from Purkinje cells—synapses increase in strength when active during *transient* decreases in the ongoing Purkinje input and decrease in strength when active during strong Purkinje input; (2) inhibition of climbing fibers by cerebellar output helps regulate spontaneous climbing fiber activity to an equilibrium level at which LTD and LTP balance and the expected net change in *gr* → *Pkj* synapses is zero; (3) interactions between the populations of granule and Golgi cells help convert temporally constant mossy fiber inputs into time-varying activation of different granule cells at different times during a stimulus (a CS); and (4) these same interactions confer a sensitivity to noise such that the across-trials consistency of the granule cells activated by a CS first increases and then gradually fades during the stimulus. This variation in across-trials consistency influences learning and in part mediates the ISI function.

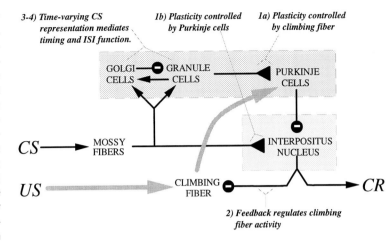

cortex may depend on the amount of training. Early in training, before the induction of significant plasticity at mf → nuc synapses, a lesion might abolish the expression of CRs, whereas later in training, lesions should spare retention of (improperly timed) CRs. This training-dependent distribution of plasticity may in part explain the variable degrees of retention that have been seen following lesions of the cerebellar cortex.

The relative distribution of plasticity between the cerebellar cortex and nucleus may be sensitive to the type of training as well. Starting with Marr's (1969) notion that mossy fiber inputs to the large population of granule cells is an expansion recoder, that is, a device that increases discriminability of different stimuli, we can view the populations of mossy fibers and of granule cells as providing relatively coarse and fine CS representations, respectively. Fairly similar mossy fiber inputs would give rise to less similar granule cell representations. Thus, we can consider a functional stimulus representation unit as being comprised of a small number of mf → nuc synapses and a much larger population of *gr* → *Pkj* subsets that can be active at the same time. As the coarser, lower resolution representation, the mf → nuc synapses might be activated by a relatively wide range of stimuli, for example, tones. The higher resolution representation provided by the subsets of *gr* → *Pkj* synapses might code a more restricted range of auditory frequencies as well as different temporal segments of the same auditory stimulus.

Given this arrangement, the strength of mf → nuc synapses may reflect the average CR coded by each *gr* → *Pkj* synapse, weighted by its frequency of activation. This follows from the notion that mf → nuc synapses increase in strength when they are coactive with a subset of *gr* → *Pkj* synapses that code for a CR. Conversely, mf → nuc synapses should decrease in strength when coactive with a subset of *gr* → *Pkj* synapses that do not produce a decrease in Purkinje cell activity. This suggests that each *gr* → *Pkj* synapse competes with its cohorts to make the strength of the mf → nuc synapses consistent with the response that it encodes: *gr* → *Pkj* synapses that have undergone LTD act to make the mf → nuc synapses stronger, and synapses that have undergone LTP act to make mf → nuc synapses weaker.

In this view, the amount of plasticity at mf → nuc synapses could depend on the type of training. Standard delay conditioning, using a single CS and a fixed ISI, may represent an example where a limited number of *gr* → *Pkj* subsets code for a CR and would lead to a moderate amount of plasticity at mf → nuc synapses—enough to allow for the expression of small CRs following removal of the cerebellar cortex. In contrast, training with several auditory CSs, or training with a single CS using several ISIs, could contribute to more plasticity in the nucleus because a CR is required under a larger number of circumstances—that is, more $gr \rightarrow Pkj$ subsets code for a CR. Consequently, after a lesion of the cerebellar cortex, a single CS should elicit a relatively robust but still inappropriately timed CR.

Our assumptions also predict that the relative distribution of plasticity varies throughout extinction training and that the distribution of synaptic weights following acquisition and extinction may be quite different than before training. During extinction training, the previously learned decrease in Purkinje cell activity during the CS is reversed by the induction of LTP at CS-activated $gr \rightarrow Pkj$ synapses. As with acquisition, our assumptions allow that CRs could be extinguished by the restoration of ongoing Purkinje activity during the CS, before a significant decrease in the strength of $mf \rightarrow nuc$ synapses. Although further extinction training my also reverse the LTP_{nuc} at $mf \rightarrow nuc$ synapses, the model predicts that a lesion of the cerebellar cortex might unmask CRs (albeit poorly timed) at the point in extinction when CRs initially fail to occur.

SAVINGS

The interactions we propose between two sites of plasticity offer possible explanations for two phenomena generally referred to as savings. One form of savings is the commonly observed finding that reacquisition following extinction occurs more rapidly than initial acquisition. Because our assumptions allow that a residual portion of the increased $mf \rightarrow nuc$ synaptic strength produced by acquisition may persist following extinction training, subsequent reacquisition would occur relatively rapidly as LTD develops at $gr \rightarrow Pkj$ synapses. At some level of abstraction, this reasoning is similar to the two-stage model of acquisition and extinction proposed by Kehoe (1988). However, unlike Kehoe's model, the present model predicts that with extended extinction training, the amount of savings should be reduced as $mf \rightarrow nuc$ synapses decrease in strength as the result of repeated activation in mossy fiber collaterals at a time when the nucleus cell is inhibited.

This putative mechanism may also account in part for a second savings phenomenon in which training to one CS can facilitate acquisition of responses to a second CS. For this mechanism to apply, there must be a degree of overlap in the mossy fiber representations of the two CSs. Although there may be little to no behavioral generalization between the stimuli owing to virtually nonoverlapping granule cell representations, the potential for savings in acquisition to the second CS following training to the first should parallel the extent of the shared mossy fiber representation. This account predicts that following training to the first CS, a lesion of the cerebellar cortex should unmask a generalized response to the second CS. Although previous studies have shown that stimulus generalization increases somewhat following a cerebellar cortex lesion (Perrett et al. 1993; S. Garcia and M.D. Mauk, unpubl.), this increase may not be sufficient to explain all savings of this form. For example, it may not explain savings to a visual CS following training to an auditory CS. Our ideas also do not explain the savings seen with training to one eye following prior training to the other eye. An explanation of these observations appears to require additional mechanisms.

RESPONSE TIMING AND THE ISI FUNCTION

Our explanations for both the ISI function and for response timing arise almost entirely from assumptions about how the CS representation in the cerebellum changes during the stimulus. We have extended Marr's (1969) original idea—that stimuli are represented by the subsets of granule cells they make active—by suggesting that interactions between the populations of granule and Golgi cells activate different subsets of granule cells throughout a stimulus. Results from recent large-scale computer simulations of the cerebellar cortex demonstrate that, in principle, networks with an organization similar to the cerebellum produce time-varying representations from tonic, time-invariant mossy fiber inputs (Buonomano and Mauk 1994). These simulations also illustrated a possible explanation for the anticipatory nature of CRs—that is that responses begin before US onset. Because the cascade from one active subset of $gr \rightarrow Pkj$ synapses to the next is continuous rather than discrete, the overlap between the currently active subset and the previously reinforced subset will increase gradually as the reinforced time approaches. As the similarity increases, Purkinje cell activity will decrease, and thus the likelihood of a response will grow, producing an anticipatory CR.

The time-varying representation of the CS generated by the cerebellar cortex may come at a cost: a sensitivity to noise that gradually degrades the across-trials consistency of granule cell activity as the CS duration increases. As we noted previously, the ISI function may arise from a rapid increase then gradual decline in the consistency from one trial to the next in the subsets of granule cells activated by the CS.

The present ideas point to the importance of characterizing the CS-related firing patterns of granule cells over many trials. Unfortunately, granule cells are small and densely packed, which makes unit recordings from awake animals extremely difficult. We know of no published unit recordings of cerebellar granule cells during Pavlovian conditioning or even during presentation of CS-like stimuli. Given the statistical and distributed nature of the CS representation we propose, empirical tests would require recording single unit responses to a large number of CS presentations from a large number of granule cells. However, if such recordings could be obtained, the peristimulus histograms should show peaks and/or valleys during the CS and different cells should have these peaks and valleys in different times. This distribution would provide the temporal discrimination for response timing. Furthermore, the overall likelihood of observing these peaks and valleys during segments of the CS, in each of a large number of cells, should be related to across-trials consistency and thus should parallel the time course of the ISI function.

A number of neural-like models that involve time-varying CS representations have been proposed to account for the timing of CRs. For example, Moore and colleagues (1986) presented a tapped delay line model in which the CS activates a serial sequence of neurons, with each synapse producing a delay. These delay lines, combined with an eligibility period for plasticity, provide an account for response timing and for the ISI function (Moore et al. 1986, 1989; Moore and Blazis 1989). Bullock and colleagues (1994) presented a cerebellar model adapted from a timing hypothesis originally proposed to occur in the hippocampus (Grossberg and Schmajuk 1989) in which response timing is derived from an array of CS-activated elements with different time constants. In a similar fashion, Gluck and colleagues (1990) proposed a time-varying CS representation comprised of CS elements that oscillate at different frequencies and phases.

Clearly, there are many ways in which neural activity related to a CS could vary over time. When combined with an activity-dependent form of synaptic plasticity, such a CS representation could generate CRs with the appropriate timing. However, the key question is how the nervous system generates such a time-varying representation. Rather than building in time-varying activity with assumptions about delay-lines or about arrays of cells or units with different temporal properties (membrane or synaptic time constants, oscillations at different frequencies, etc.), our approach has been to ask if time-varying activity can emerge naturally from networks configured like the cerebellar cortex (see Buonomano and Mauk 1994).

We have used the same approach in attempting to account for the ISI function. We have asked whether, like CR timing, a mechanism for the ISI function can arise as a network property of the anatomy and physiology of the cerebellar cortex. This is in contrast to previous models in which the ISI function has generally been obtained by building in an eligibility parameter for plasticity that explicitly mirrors or reproduces the time course of the ISI function (Grossberg and Schmajuk 1989; Moore and Blazis 1989; Moore et al. 1989). Our account for the ISI function is obtained without the addition of noncerebellar features and without imposing parameters explicitly calculated to produce the ISI function. This mechanism also permits eligibility for plasticity to be related to the ongoing CS-related spike activity (in the $gr \rightarrow Pkj$ synapses). For example, climbing fiber inputs can simply modify the strength of those $gr \rightarrow Pkj$ synapses that are active. There is no need to assume that climbing fiber inputs interact with an eligibility signal whose time course *explicitly mirrors and directly produces the ISI function*, as is the case with many models.

PERMANENT LESIONS OF THE CEREBELLAR CORTEX

Because the cerebellar nucleus is assumed to be necessary for the expression of CRs, the ability of the model to explain findings that interpositus nucleus lesions abolish CRs is unsurprising. On the other hand, predictions regarding the effects of cerebellar cortex and IO lesions are derived from the interactions of several assumptions of the model.

A firm prediction of the model is that the timing of any CRs observed following lesions of the cerebellar cortex should be disrupted in a characteristic manner. Postlesion responses should show short latencies to onset and to peak that are, at most, minimally sensitive to the ISI used in training (that is, to the extent that the strength of the mf → nuc synapses can influence the time required to reach threshold for response production). These expectations have been confirmed by Perrett et al. (1993). This study used within-subject comparisons in which animals were trained with two CSs, each reinforced at a different ISI. This training produced differently timed responses to the two CSs. Following lesions of cerebellar cortex, latencies to onset and to peak were short, and the differential timing observed prior to the lesion was abolished; i.e., the postlesion CRs displayed the same, short latency independent of their prelesion timing. These findings are consistent with the proposal that under most circumstances, CRs in a well-trained animal are mediated by plasticity in both the cerebellar cortex and nucleus. Disruption of CR timing is consistent with the assumption that the cerebellar cortex generates a time-varying representation of the CS in the cortex and that gr → Pkj synapses active at the time of US occurrence undergo plasticity. Retention of improperly timed CRs following cortex lesions is consistent with the idea that plasticity also occurs at the mf → nuc synapses.

In the previous sections we outlined reasons why the retention of CRs following lesions of the cerebellar cortex should depend in complex ways on the amount and type of prelesion training. This leaves open the possibility that cerebellar cortex lesions will abolish responses early in training and will spare improperly timed responses after additional CS + US training. We believe that this possibility may allow a better understanding of apparent inconsistencies in the effects of cerebellar cortex lesions on the retention of CRs that have been reported. The original report on the effects of cerebellar cortex lesions (McCormick and Thompson 1984), as well as many other subsequent studies (Perrett et al. 1993; Perrett and Mauk 1995; L.S. Garcia and M.D. Mauk, unpubl.), shows retention of CRs following cerebellar cortex lesions. However, in other studies (Yeo et al. 1985b; Yeo and Hardiman 1992), similar lesions abolished the retention of previously acquired eyelid responses. As has been argued by Lavond and Steinmetz (1989), one explanation of these differences suggests that the abolished responses reflect unintended damage to the underlying interpositus nucleus that cannot be detected by histological analysis. Although this is possible, our model allows a different perspective that is amenable to experimental evaluation—that differences in the amount and type of training that an animal has had before the lesion can explain the variable lesion effects that have been reported. For example, with modest amounts of training, CRs may be mediated largely by LTD in the cortex, whereas with extended training, CRs may be produced by plasticity in both the cerebellar cortex and interpositus nucleus. The effects of a cerebellar cortex lesion on the retention of CRs would be quite different for these two circumstances (Mauk 1997).

Another strong prediction of the model is that cerebellar cortex lesions should prevent both the extinction of previously learned responses and the acquisition of new CRs. Although, within our assumptions, one site of plasticity (mf → nuc synapses) is spared by such a lesion, these predictions arise from the idea that plasticity at this remaining site requires either Purkinje cell input (for LTD_{nuc}) or *transient release* from Purkinje cell input (for LTP_{nuc}). If this is true, the amplitude of the response seen initially after the lesion should not change with further acquisition or extinction training. A recent experiment has shown that lesions of the cerebellar cortex prevent extinction of eyelid CRs over extensive CS-alone training (Perrett and Mauk 1995). This study and a previous study (Perrett et al. 1993) found that lesions must include the anterior lobe of the cerebellar cortex to affect both the timing of the CRs and extinction. In addition, preliminary results of Garcia and Mauk (1995) also support the prediction that cerebellar cortex lesions prevent acquisition of new responses. This study also confirmed that damage to the anterior lobe is required to produce the effect. Together, these studies are consistent with the hypotheses that (1) CRs in well-trained animals are mediated by plasticity in both the cerebellar cortex and nucleus, (2) both a time-varying CS representation and synaptic plasticity in the cerebellar cortex are responsible for CR timing, and (3) following lesions of the cerebellar cortex, the ability of mf → nuc synapses to undergo plasticity is greatly diminished.

In contrast, two previous studies showed that cerebellar cortex lesions produce deficits in, but do not completely abolish, acquisition (Lavond et al. 1987; Lavond and Steinmetz 1989). The histological reconstructions presented in these papers indicate that considerable parts of the anterior lobe were spared in these experiments. We believe that an appropriate test must focus on the involvement of the anterior lobe and at the same time be very careful to provide assurances that there is no damage to structures known to be necessary for the expression of responses such as the interpositus nucleus or the middle or superior cerebellar peduncles.

GENETIC MANIPULATIONS OF CEREBELLAR CORTEX

Recently, mice with genetic mutations that affect the cerebellum have been used to address cerebellar involvement in eyelid conditioning. In one such study, Chen et al. (1996) report that *pcd* mice, which lose all Purkinje cells in the third to fourth week of development (functionally disconnecting the cerebellar cortex from the cerebellar nuclei), show some acquisition of eyelid CRs, although acquisition is "impaired profoundly" (p. 2834). The authors also report that these CRs extinguish over 4 days of CS-alone presentations. The authors conclude that "This finding suggests that the Purkinje cells and thus the entire cerebellar cortex (because the Purkinje cells are the sole efferent neurons of the cortex) are not necessary for the extinction of eyeblink CRs" (p. 2835). Another study, using mice lacking one isoform of metabotropic glutamate receptors (mGluR$_1$), used in vitro cerebellar slice to show a lack of LTD at $gr \rightarrow Pkj$ synapses. This deficit in LTD was accompanied by an impairment, but not complete abolition, of the ability to acquire conditioned eyelid responses (Aiba et al. 1994; see also Chen et al. 1995).

Although these results appear to contradict previous lesion results and contradict predictions of the present model, their implications for cerebellar mechanisms of eyelid conditioning are not yet entirely clear owing to a number of unresolved issues that make drawing conclusions from the findings difficult. For example, Chen et al. (1996) acknowledge a number of ways in which the *pcd* mutant mice differ from wild type. In addition to lacking Purkinje cells, they have smaller IO cell bodies, show abnormal neurochemistry in the interpositus nucleus (Triarhou and Ghetti 1991; Goldowitz and Eisenman 1992; Godlett et al. 1992), show ataxia, and fail to learn the hidden platform problem in the Morris water maze (suggesting abnormal development of the hippocampus). These specific abnormalities exemplify the major conceptual problems with most learning studies using mutant mice, namely, the potential for widespread abnormalities (both obvious and subtle) and the corresponding potential for compensatory mechanisms to be engaged during development or during learning. Thus, to conclude that learning abilities observed in mutant animals demonstrate that the most obviously affected brain systems are not involved in such learning in normal animals is problematic. To make such a claim, given the potential for the many confounds described above, it is absolutely essential to show that any residual learning (or extinction) in mutant animals is mediated by the same processes involved in normal animals.

Taking the Chen et al. (1996) observation of extinction in mice with no Purkinje cells as an example, we note first the absence of controls critical for determining whether the loss of responding during 4 days of CS-alone training was attributable to extinction rather than forgetting. Given the variety of structural and physiological abnormalities in these mice, there is no way of knowing whether or not retention is normal or whether forgetting is more rapid than in wild-type mice. Second, if CRs do indeed extinguish in these animals, it is important to know whether the extinction is mediated through a compensatory mechanism or the mechanism that operates in normal mice. It is also important to determine whether the impaired, but still present, acquisition seen in the mutant mice depends on the cerebellum. Unfortunately, the key question of whether lesions of the interpositus nucleus in the mutant mice abolish the small CRs that are acquired was not evaluated. Without such controls, it seems quite possible that the (impaired) acquisition that is seen in mutant mice may be attributable to compensatory, even noncerebellar, mechanisms.

PERMANENT LESIONS OF THE IO

The assumptions of the model do not permit a firm prediction concerning the retention of CRs following a lesion of the IO. The uncertainty stems from the unknown effects of removing spontaneous climbing fiber activity. Without it, $gr \rightarrow Pkj$ synapses would undergo LTP each time they are active, and, presumably, this would lead to a reduction or abolition of previously trained CRs. The surest way of activating these synapses is to present the CS. However, they will also be active to a lesser degree at other times, and in both cases this should induce LTP and diminish the probability of

CRs. Thus, the interval between the IO lesion and testing for retention of CRs could affect the ability of the CS to elicit a CR. Relatively long lesion-to-test intervals should reduce the likelihood of observing postlesion CRs, as should recovery in environments that increase the probability that the $gr \rightarrow Pkj$ synapses coding the pretrained CS are active during the recovery period (e.g., a noisy home case environment). Various combinations of these factors could be responsible for the differences in the results of two studies examining IO lesions, one of which reported an abolition of CRs (Yeo et al. 1986) and the other of which reported *initial* retention of CRs (McCormick et al. 1985). In addition, the lesions reported by Yeo et al. were relatively large compared to those reported by McCormick et al. and may therefore have abolished CRs as a result of more widespread disruption of cerebellar function. However, our model makes the firm prediction that any responses retained after an IO lesion will extinguish with additional CS + US training, which was observed in the McCormick et al. study.

REVERSIBLE LESIONS OF THE INTERPOSITUS NUCLEUS AND IO

Several studies have used reversible lesions using local anesthetics such as lidocaine or specific transmitter agonists or antagonists to analyze cerebellar involvement in eyelid conditioning. Mamounas et al. (1987) reported that infusion of the $GABA_A$ antagonists into the interpositus nucleus abolishes CRs for 10-20 trials, followed by a gradual return to preinfusion baseline. Because at least some of the animals received extensive preinfusion training, our assumptions predict significant plasticity at $mf \rightarrow nuc$ synapses, and thus we cannot explain the initial abolition of CRs. To the degree that the drug effect simulated removal of the cerebellar cortex, we would expect that response timing during the recovery phase would be disrupted. Unfortunately, response timing was not investigated in this study.

In an elegant and important series of studies, Welsh and Harvey (1991) and Krupa et al. (1993) assessed the effects of inactivating the interpositus nucleus with lidocaine during acquisition training. Welsh and Harvey observed significant retention of CRs in post-training tests during which the interpositus nucleus was free of lidocaine. Using a similar protocol, Krupa et al. found no retention of responses following a training phase during which the nucleus cells were hyperpolarized with the GABA agonist muscimol.

We suggest that both observations are consistent with the present model and that the different modes of action of lidocaine and muscimol led to the different experimental outcomes. We believe that the induction of LTD at $gr \rightarrow Pkj$ synapses and the acquired decreases in Purkinje activity proceeded normally in both experiments, perhaps even more robustly than normal without response-related inhibition of climbing fibers. However, because lidocaine blocks both pre- and postsynaptic action potentials, no activity-dependent plasticity should occur in the interpositus nucleus during its infusion. The subsequent expression of CRs in the absence of lidocaine would reflect training-induced plasticity in the cerebellar cortex sufficient to generate CRs, even though the strength of nucleus synapses remained unchanged. In contrast, our assumptions predict that training during infusion of muscimol in the interpositus would promote extinction-like processes at the $mf \rightarrow nuc$ synapses, that is, a decrease in synaptic strength owing to activation of $mf \rightarrow nuc$ synapses during conditions that mimic strong Purkinje cell input (i.e., muscimol). Without knowing how this mix of functionally opposing synaptic changes in the cortex and nucleus would affect cerebellar output, it is difficult to predict the degree of retention that would be seen. Overall, our assumptions are qualitatively consistent with the immediate appearance of CRs after training using an infusion of lidocaine and with a more delayed appearance of CRs observed following training with muscimol infused into the interpositus nucleus.

More recently, Nordholm et al. (1993) demonstrated that infusing lidocaine into dorsal regions of interpositus nucleus prevents acquisition of CRs, whereas infusing it into more ventral regions does not (even though it still blocks the expression of CRs). At face value, these data contradict the hypothesis that conditioning is driven by training-induced plasticity in the cerebellar cortex and, as such, would require a major modification of the present model. However, one important possibility that was not excluded in these experiments is that lidocaine infused into the dorsal regions blocked climbing fiber or mossy fiber inputs to the cerebellar cortex.

Finally, one study has shown that cooling the IO produces increases in Purkinje cell activity that reverse with the termination of cooling (Montarolo et al. 1982). These results are consistent with the proposed self-regulating modulation of climbing fiber activity. Without appropriate levels of spontaneous climbing fiber activity (e.g., during cooling), the induction of LTP at $gr \rightarrow Pkj$ synapses, unopposed by LTD, would lead to increased activity of Purkinje cells.

PREDICTIONS OF THE MODEL RELATED TO REVERSIBLE LESIONS

In general, the use of reversible lesions is fertile ground for further investigating cerebellar involvement in motor learning such as eyelid conditioning. One important aspect of our model is the large number of empirically testable predictions that it makes regarding the effects of various reversible lesion experiments.

Our assumptions suggest that the extinction of CRs should be blocked by any reversible lesion that disrupts the inhibitory pathway from cerebellar nuclei to the climbing fibers in the IO. Under such circumstances, the presentation of CS-alone extinction trials would not be expected to induce LTP at the CS-activated $gr \rightarrow Pkj$ synapses because the response-related inhibitory output from the cerebellum to the IO is required to reduce spontaneous climbing fiber activity below the equilibrium level. According to this thinking, extinction should not occur when the CS-alone extinction trials are delivered during the following reversible lesions: (1) any infusion that reversibly blocks the ability of interpositus nucleus output to inhibit climbing fiber activity, such as lidocaine or muscimol in the interpositus nucleus or lidocaine in the superior cerebellar peduncle and (2) infusion of GABA antagonists into the IO, which should also disrupt inhibition of climbing fibers initiated by cerebellar output.

In addition, the model predicts that infusions that block extinction should also abolish the shift in response timing seen when training begins with one ISI and then is shifted to a second ISI. For example, Coleman and Gormezano (1971) and also Millenson et al. (1977) initially trained animals at one ISI until robust CRs were observed; additional training was then conducted at a different ISI. The authors showed that changing the ISI produced a gradual shift in response timing; latencies to onset and to peak gradually became appropriate for the second ISI. Given our assumptions, the timing shift seen in this protocol involves the extinction of the response timed appropriately for the first ISI and the simultaneous acquisition of a response with timing appropriate for the second ISI. With extinction disrupted, shifting to the new ISI should result in acquisition of a response timed appropriately for the second ISI without extinguishing the first, producing double-peaked responses as has been observed by D.J. Krupa, J. Weng, and R.F. Thompson (unpubl.).

It is also clear from the assumptions of the model that reversible lesions that prevent activity in Purkinje cells or disrupt their inhibitory input to nucleus cells should also disrupt the timing of CRs. Moreover, to the extent that rebound excitation contributes to activation of nucleus cells, such manipulations may decrease the amplitude of CRs as well. These effects should be produced by the following reversible lesion protocols: (1) GABA antagonists are infused into the interpositus nucleus to block the inhibition of nucleus cells by Purkinje cells. Unless these antagonist infusions involve complex interactions with local inhibitory interneurons, the effects of such infusions should be the equivalent of a reversible disconnection of the cerebellar cortex from the nucleus. (2) Local anesthetics or the GABA agonist muscimol is infused into the anterior lobe of the cerebellar cortex. With no inhibitory input, the nucleus cells could become active as soon as the CS input arrives via the mossy fibers. This protocol should reduce the amplitude of the CRs by removing the contribution of the cortex to CR generation (i.e., by removing rebound excitation in the nucleus cells).

In addition, the infusion of a GABA antagonist such as picrotoxin into the anterior lobe of the cerebellar cortex should abolish the expression of CRs. Without inhibitory input from stellate and basket cells, the decrease in excitatory synaptic input that occurs during the CS owing to training and the induction of LTD may not be sufficient to produce a robust decrease in the activity of the Purkinje cell. If so, the ongoing Purkinje cell inhibition of nucleus cells should prevent the expression of the already learned responses.

OTHER CONDITIONING PHENOMENA

Our assumptions also provide fairly simple accounts for more complex aspects of conditioning such as blocking, conditioned inhibition, differential conditioning, and trace conditioning. The Kamin blocking effect is the finding that the amount of conditioning produced by a CS + US pairing is diminished when the target CS appears in compound with a second CS that has previously been trained with that US, compared with occasions in which the target is trained in compound with a second CS that has not been paired with the US (Kamin 1969). A well-controlled design would look like the following:

	Phase 1	Phase 2	Test
Blocking	CS1 → US	(CS1 + CS2) → US	responding to CS2
Control	CS3 → US	(CS1 + CS2) → US	

The blocking effect is the finding that during the test phase, the level of conditioned responding to CS2 is lower in the blocking group than in the control group. Thus, prior training to CS1 blocked conditioning to CS2. The phenomenon of blocking is quite important because it demonstrates clearly that although the pairing of a CS and US is necessary for conditioning to occur, it is not a sufficient condition. Why this should be so is a question that has motivated a considerable amount of experimental and theoretical work in the Pavlovian conditioning literature (Rescorla and Wagner 1972; Makintosh 1975; Wagner 1976, 1979, 1981; Pearce and Hall 1980).

Successful theoretical accounts for blocking have been obtained in a number of models that assume that an expected US is less effective than an unexpected US (e.g., Wagner and Rescorla 1972; Wagner 1976, 1981; Sutton and Barto 1981). Moreover, a number of authors have noted that the inhibition of climbing fibers by cerebellar output represents a potential biological implementation of this theoretical construct (e.g., Donegan et al. 1989). An account of blocking in the present model is obtained in this fashion and is presented only to highlight aspects that may be unique to our model. Essentially, the prior training to CS1 in the blocking group produces a non-zero level of responding to the CS1 + CS2 compound in phase two. As such, less learning occurs during the compound because the US-evoked climbing fiber response is somewhat diminished, resulting in less conditioning to the CS2.

A number of experimentally testable predictions arise from the assumption that blocking is produced by increased inhibition of climbing fibers during the compound phase. The blocking effect should be sensitive to the reversible lesions that prevent cerebellar output from inhibiting climbing fiber activity during the second phase. During the phase 2 compound training, lidocaine or muscimol infused in the cerebellar nucleus or a GABA antagonist such as picrotoxin infused in the IO should prevent blocking. These predictions are consistent with preliminary findings of J.J. Kim, D.J. Krupa, and R.F. Thompson (unpubl.) who report that intra-olivary infusions of picrotoxin prevent blocking in rabbit eyelid conditioning. Moreover, given mechanisms of temporal discrimination in the cerebellar cortex that we hypothesize, the present model makes the prediction that blocking may show a degree of dependency on the ISI. Specifically, if the ISI in phase 1 is not the same as in phase 2, then blocking should be attenuated. This prediction simply acknowledges the expectation that the CR-related inhibition of climbing fibers should show ISI-dependent timing similar to the CRs themselves. To our knowledge, this prediction has not yet been tested experimentally.

An alternative way in which the effectiveness of the US might be reduced could be through the desensitization of the trigeminal nerve produced by the closure of the eyelid during the CR (Pellegrini and Evinger 1995). Although such effects may contribute to the blocking effect, we would predict that if the facial motor nucleus was inactivated by lidocaine during the compound phase of the training phase (thereby preventing CS1 from eliciting a CR), a substantial amount of blocking would be observed in the test phase in the absence of lidocaine. That is, responding to CS2 would still be significantly lower in such a blocking group, compared with the control group.

Our assumptions suggest a parallel account for conditioned inhibition, a phenomenon that was first reported by Pavlov (1927) and is procedurally quite similar to blocking. The main difference is that the compound (CS1 + CS2) is not reinforced in the second phase. An example of a procedure for developing and demonstrating conditioned inhibition is

	Phase 1	Phase 2	Test
Conditioned inhibition	CS1 → US CS3 → US	(CS1 + CS2) → no US	
Control 1	CS3 → US	(CS1 + CS2) → no US	CS3 alone vs. (CS2 + CS3)
Control 2	CS1 → US CS3 → US	CS1 → no US CS2 → no US	

min afterward. LTD was induced by one of two protocols: (1) Stimulus intensity was raised and the amplifier was set to current-clamp mode; in this case, PF-excitatory postsynaptic potentials (EPSPs) were evoked at 1 Hz for 1–2 min; or (2) the Purkinje cell was voltage clamped and stimulus intensity was kept constant while PFs were activated in brief bursts of stimuli (two to five pulses per train, at a frequency of 10–50 Hz) applied every second for 1–2 min. Peak amplitudes of the electrical responses were measured with Pulse (HEKA) and transferred to Igor Pro (Wavemetrics, Lake Oswego, OR) for further analysis. LTD was defined as an irreversible decrease of >20% in PF response amplitude.

FLUOROMETRIC CALCIUM IMAGING

Changes in the intracellular calcium concentration were recorded with a video-rate confocal laser-scanning system (Odyssey, Noran, Middletown, WI) that was mounted on an upright microscope (Axioplan, Zeiss; Eilers et al. 1995b) equipped with water-immersion objectives (Achroplan, Zeiss, 40×, NA 0.75 and 63×, NA 0.9). Full-frame images usually were collected at video rate (30 Hz) and stored on an optical disc recorder (TQ-2026F, Panasonic, Secaucus, NJ). Images were processed off-line with Image-1 software (Universal Image Co., West Chester, PA). In some experiments images were digitized on-line (MVP-AT frame-grabber board, Matrox, Dorval, Quebec, Canada), integrated over periods of 500 msec (Image 1) and stored on the hard disk of a personal computer. In either case, fluorescence data were reduced to graphic form by measuring fluorescence from areas of interest and calculating the background-corrected increase in fluorescence divided by the prestimulus fluorescence ($\Delta F/F$) using Igor Pro. Such ratios yield a parameter that is independent of optical path length and dye concentration and, at a given resting Ca^{2+} level, is a unique function of intracellular Ca^{2+} concentration (e.g., Neher and Augustine 1992). The laser was shuttered between recording episodes to minimize phototoxic damage. Experiments showing any signs of such damage (e.g., irreversible changes in fluorescence or increases in Purkinje cell holding current) were not accepted for analysis.

Results

We began our study of PF-LTD by defining experimental conditions that allowed us to reliably evoke PF-LTD and then used imaging techniques to visualize the Ca^{2+} signaling events associated with this form of synaptic plasticity.

REQUIREMENTS FOR INDUCTION OF LTD BY PARALLEL FIBER ACTIVITY

Stimulation of PFs at relatively low intensity and low frequency (0.1 – 0.2 Hz) produced postsynaptic electrical responses that were stable over prolonged periods of time. These conditions were therefore chosen to monitor basal PF synaptic transmission. At low stimulation intensities, increasing the stimulation frequency to 1 Hz for 1 to 2 min did not consistently depress synaptic transmission ($n = 5$; data not shown; see also Hartell 1996a). However, more vigorous stimulation of PFs was capable of eliciting LTD. We established two PF stimulation protocols that reliably produced LTD. With both stimulus paradigms, PF transmission was partially depressed immediately after the stimulus. The magnitude of the depression then increased progressively over the next 30 min and was stable for the duration of the recording (up to 2 hr). In both cases, LTD caused a reduction in the amplitude of PF-EPSCs, without any change in their time course. It is important to note that neither of these PF stimulation paradigms evoked action potentials, as monitored by recordings from the Purkinje cell body.

In the first stimulus regimen, stimulation intensity was kept constant but PFs were activated with short bursts of stimuli (two to five pulses at 10–50 Hz) at a frequency of 1 Hz for 1 to 2 min. This protocol produced a depression of PF synaptic responses in 20 out of 29 cells, with a mean decrease in EPSC amplitude to 58% ± 2% (±s.e.m.) measured 25 min after the start of the induction (Fig. 1A). In the second paradigm, LTD was induced while switching the patch clamp amplifier from voltage clamp to current clamp mode. Turning off the voltage clamp allows the Purkinje cell membrane potential to depolarize more during a PF response, thereby increasing the amount of Ca^{2+} entry evoked by a given number of active PFs (Eilers et al. 1995a). In addition, stimulus intensity was increased to recruit additional PFs. Under these conditions, activation of PFs at 1 Hz for 1–2 min induced a long-lasting depression of PF responses in 8 out of 10 cells, with a mean decrease to 56% ± 1% in the amplitude of PF-EPSCs (Fig. 1B).

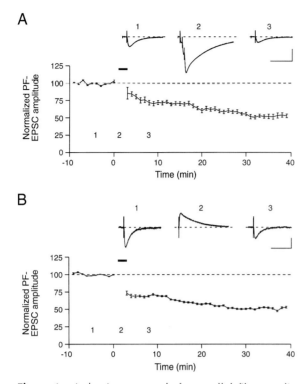

Figure 1: Induction protocols for parallel fiber mediated LTD. Two induction protocols were used, both of which produced LTD that developed over a similar time course. (*A*) The first protocol consisted of short bursts of PF stimulation (two to five stimuli at 10–50 Hz), which were repeated at 1 Hz for 1–2 min. This stimulation protocol was given in voltage-clamp mode. For each experiment the responses were binned to 1 min, and the mean and s.e.m. values were calculated within the two experimental groups. Horizontal bars indicate the induction periods. Experiments were counted as failures when the reduction in synaptic response was <20%. EPSCs taken from a representative experiment are shown (*upper right*). The EPSCs were averaged over 2 min at the indicated time points. Scale bar, 100 pA/100 msec for EPSCs 1 and 3 and 100 pA/200 msec for EPSC 2. (*B*) The second protocol utilized single-shock stimulations of increased intensity, which were applied in current-clamp mode. Stimulation was repeated at 1 Hz for 1–2 min. Electrical responses taken from a representative experiment are shown (*upper right*). The responses were averaged over 2 min at the indicated time points. Scale bar, 20 pA/100 msec for response 1 and 3 and 2 mV/200 msec for response 2.

IMAGING PURKINJE CELL CALCIUM SIGNALS DURING INDUCTION OF LTD

The PF stimulation patterns that we have found to produce LTD, that is, burst stimulation and/or strong PF stimulation, are known to produce local and transient elevations in dendritic $[Ca^{2+}]_i$ (Miyakawa et al. 1992; Eilers et al. 1995a; Hartell 1996a). However, previous studies have left unresolved the questions of whether these dendritic Ca^{2+} signals remain spatially restricted during repetitive activity of PFs and whether local Ca^{2+} signals are indeed sufficient to produce LTD. We have addressed these questions by measuring postsynaptic Ca^{2+} concentration changes during the induction of PF-LTD.

To define the spatial range of postsynaptic Ca^{2+} signaling during PF-LTD induction, we combined whole-cell patch clamp recordings of Purkinje cell electrical responses with simultaneous imaging of changes in intracellular Ca^{2+} concentration ($[Ca^{2+}]_i$). The following data were obtained from 10 experiments in which simultaneous patch-clamp and imaging recordings were acquired from single Purkinje cells. To measure $[Ca^{2+}]_i$, Purkinje cells were filled with a Ca^{2+} indicator dye, such as Oregon BAPTA Green-1, which was delivered by the patch pipette. An image of a Purkinje cell filled with a fluorescent Ca^{2+} indicator dye is shown in Figure 2. This method of delivery causes a gradual increase in dye concentration as the dye diffuses from the patch pipette throughout the dendrites (Pusch and Neher 1988; Rexhausen 1992). We began our measurements at least 20 min after starting to introduce the dye; by this time the dye concentration should have been well equilibrated in the dendritic compartments under study and has been estimated to reach 40%–45% of the concentration

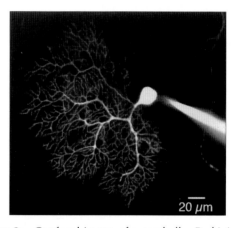

Figure 2: Confocal image of a cerebellar Purkinje cell from an 18-day-old rat. The cell was filled with the calcium indicator dye Calcium Green 1 via the somatic patch pipette (*right*). Afferent PFs were activated with an extracellular stimulation electrode (not shown) that was located in the molecular layer.

of dye within the pipette (Rexhausen 1992). Dye concentration should have increased no more than 5% during the course of LTD induction, so that any images acquired during this time are unlikely to be distorted significantly by time-dependent changes in dye concentration.

IDENTIFICATION OF THE SITE OF ACTIVE PF SYNAPSES

The dendritic tree of the Purkinje cell receives glutamatergic excitatory synaptic inputs from many thousands of PFs. Thus, one of the technical challenges in performing optical imaging of postsynaptic Ca^{2+} signaling during PF-LTD is to determine the location of active PF synapses. PF synapses are oriented perpendicular to the plane of the Purkinje cell dendritic tree; in the parasagittal slices utilized in our experiments, PFs are perpendicular to the horizontal surface of the slice. Such an arrangement makes it possible to activate a variable number of PF synapses simply by placing an extracellular stimulation pipette on the surface of the slice. In such conditions, it was fairly easy to identify the site of PF activity by monitoring the Ca^{2+} transient associated with the PF stimulation.

A short train of PF responses evoked by burst stimulation reliably produced localized dendritic Ca^{2+} signals (Fig. 3C,D), even at stimulus intensities too weak to generate dendritic Ca^{2+} signals during single stimuli (Fig. 3A,B). These Ca signals were prevented by CNQX, suggesting that burst stimuli cause summation of local, AMPA receptor-mediated depolarizations that activate voltage-gated calcium channels. The Ca^{2+} response to a single burst stimulus was therefore used to identify the site of active PF synapses. This stimulus did not depress synaptic transmission (not shown). Ca^{2+} signals associated with PF activation could be de-

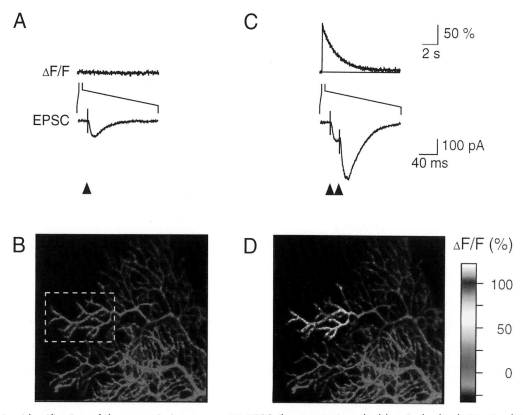

Figure 3: Identification of the synaptic input area. (A) EPSC (lower trace) evoked by single-shock PF stimulation. No accompanying change in the Ca^{2+}-dependent dendritic fluorescence was detected (upper trace). The arrowhead denotes the time of PF stimulation. (B) Pseudocolor image of the fluorescence during this single-shock PF stimulation. The dashed box indicates the region in which the dendritic fluorescence was averaged for the traces shown in A, C, and Fig. 4A. (C) EPSC and Ca^{2+}-dependent fluorescence transient evoked by a brief burst stimulation. (D) Pseudocolor image showing the spatial extent of the Ca^{2+} signal during the burst stimulation.

tected 20–30 min after establishing the whole-cell configuration. Excitatory synaptic currents produced by PF activity (PF-EPSCs) were then recorded for ~10–15 min (see below) to establish the baseline level of PF transmission. Measurements of Purkinje cell Ca^{2+} levels were begun just before applying LTD-inducing stimuli. These measurements ensured that PF responses to single stimuli were not accompanied by a Ca^{2+} transient and also served to quantify the spatiotemporal extent of the Ca^{2+} transient produced by the burst stimulation (Fig. 3C,D). Control experiments demonstrated that imaging of PF-evoked Ca^{2+} signals did not interfere with the induction of LTD (not shown).

TRANSIENT AND LOCAL ELEVATION OF DENDRITIC Ca^{2+} DURING INDUCTION OF LTD

The spatial and temporal dimensions of dendritic Ca^{2+} signals resulting from single bouts of PF activity are well characterized (Fig. 3; Denk et al. 1995; Eilers et al. 1995a, Hartell 1996a). However, the degree of localization of Ca^{2+} signals that accompany the generation of LTD by intense or repeated PF activity is not known. The high spatial and temporal resolution of our confocal imaging method allowed us to characterize PF-evoked Ca^{2+} signals, within the dendritic arbor of individual Purkinje cells, during induction of LTD.

The patterns of PF stimulation that were sufficient to induce LTD produced substantial increases in dendritic $[Ca^{2+}]_i$ that were sustained throughout the stimulus period and then decayed back to baseline. In the example shown in Figure 4, which is taken from the same experiment as Figure 3, LTD was induced by a single burst stimulus (50 Hz) given repetitively at 1 Hz for 90 sec. Following this period of stimulation, the amplitude of PF-EPSCs decreased over ~10 minutes and then stabilized at 60% of its initial amplitude. The depression in the peak amplitude of PF-EPSCs was not accompanied by a significant change in EPSC time course (Fig. 4A).

During the induction period, the dendritic Ca^{2+} signals reached a plateau level within the first three to four burst stimuli and remained at this level until the end of the stimulus train (Fig. 4C). The plateau in fluorescence intensity may have been attributable to saturation of the high-affinity indicator dye, which has a Kd of ~200 nM. Although we may not have detected the peak Ca^{2+} levels achieved during the train of stimuli, these

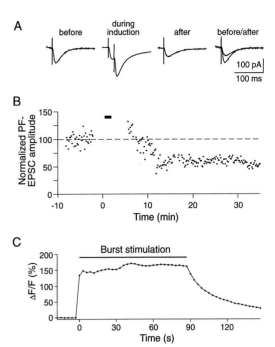

Figure 4: Elevated dendritic Ca^{2+} concentration during the induction of LTD. (A) Averages of EPSCs recorded directly before, during, and 20 min after LTD induction. The traces labeled before/after show the averaged response from before and after, superimposed for comparison. EPSC records are averages of six responses, except for the EPSC labeled during induction, which is the average of all responses during the induction period. Stimulation artifacts were reduced for clarity. (B) Time course of EPSC peak amplitudes during the LTD experiment. Values are normalized with respect to the mean baseline value. Except for during the induction period, PFs were activated with single-shock stimulations at 0.1 Hz. Individual data points are shown, with no binning. The horizontal bar indicates the induction period (corresponding to the horizontal bar in C). LTD was induced by short 50-Hz trains of PF stimulation given at 1 Hz for 90 sec. (C) Time course of the Ca^{2+}-dependent dendritic fluorescence during the induction period. For measurement, fluorescence images were averaged over a period of 500 msec and acquired at ~0.4 Hz. The dendritic Ca^{2+} signal was averaged in the same dendritic compartment as in Fig. 3B. Note the slow decay time of the fluorescence signal at the end of the induction period.

Ca^{2+} signals were clearly much larger than those produced by a single burst stimulation (cf. Fig. 3 with Figs. 4 and 5). When PF stimulation was halted, fluorescence intensity returned to its baseline level with a time constant of ~15 sec. This decay time was substantially prolonged compared to that evoked by a single burst of stimuli or a single strong stimulus (see Eilers et al. 1995a; Hartell 1996a), both of which decayed with a time constant of ~2 sec (Fig. 3C). These prolonged and large increasess in dendritic Ca^{2+} levels in Ca^{2+} levels were observed in a total of five experiments employing burst stimuli; in every case, LTD was induced by these stimuli. In contrast, in three other experiments where a train of burst stimuli ation failed to produce a prolonged buildup of dendritic Ca^{2+} or a larger Ca signal than that produced by a single burst, no depression of PF transmission was induced. These correlations indicate that the dendritic Ca^{2+} signals are necessary to induce LTD.

The repetitive stimulation of PFs during induction of LTD yielded Ca^{2+} signals (Fig. 5) that were restricted to the same dendritic compartment that was activated during single bursts of PF activity (Fig. 3). For both types of PF activity, the Ca^{2+} increase was localized to the same set of terminal dendrites and connecting parent dendrites. Thus, even during prolonged Ca^{2+} transients there was no significant spread of the signal to neighboring dendritic branches. These results indicate that very effective compartmentation mechanisms restrict the dendritic movement of Ca^{2+}. Our finding that these highly localized increases in dendritic $[Ca^{2+}]_i$ provide a sufficient Ca^{2+} signal to induce LTD provides a functional role for local PF-evoked Ca^{2+} signals and confirms the idea that the terminal dendritic branches of Purkinje cells act as functional signaling units (Eilers et al. 1995a).

Discussion

The major finding of this study is that induction of PF-LTD is associated with a transient elevation of $[Ca^{2+}]_i$ that is restricted to the spiny dendritic area activated by a beam of PFs. Although these Ca^{2+} signals persisted throughout the conditioning stimulus, they recovered to baseline values in less than a minute and did not spread to distant dendrites. These findings, combined with the observations that intracellular Ca^{2+} chelators prevent the induction of PF-LTD (Hartell 1996a), indicate that localized increases in dendritic $[Ca^{2+}]_i$ can provide a Ca^{2+} signal sufficient to induce LTD. Such local Ca^{2+} signals have been reported before, both in Purkinje cells (Denk et al. 1995; Eilers et al. 1995a; Hartell 1996a) and in hippocampal pyramidal cells (Magee et al. 1996). While it has been proposed that these signals play a role in dendritic integration, our study is the first to unambiguously assign a functional role to these signals.

The important Ca^{2+} signal for the induction of

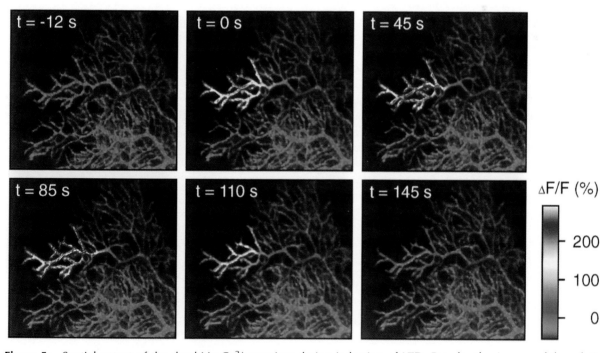

Figure 5: Spatial extent of the dendritic Ca^{2+} transient during induction of LTD. Pseudocolor images of the relative increase in Ca^{2+}-dependent fluorescence during the induction of PF–LTD. Same experiment as in Fig. 4. The indicated time points refer to the time axis in Fig. 4C.

PF-LTD thus appears to be a transient and highly localized buildup of Ca^{2+} in the vicinity of active PF synapses. Because a high-affinity Ca^{2+} indicator was used to measure postsynaptic Ca^{2+} signals, Ca^{2+} concentration at the local sites of active PFs may have been underestimated because of dye saturation. Thus, our data do not allow us to quantify the Ca^{2+} concentrations needed for the triggering of the LTD. Nonetheless, we can conclude that Ca^{2+} levels return to baseline levels rapidly (in <1 min) following the PF activity that induces LTD. Even though postsynaptic Ca^{2+} concentration rises only transiently, PF-LTD develops gradually over the subsequent 10–20 min. Thus, in terms of kinetic properties, the Ca^{2+}-dependent induction of LTD by PF activity closely resembles the induction of LTD produced by CF activation or by Purkinje cell depolarization even though the Ca^{2+} signals produced by Purkinje cell depolarization or by CF activity are more spatially widespread (Konnerth et al. 1992). At present it is not clear whether these various LTD-inducing stimuli yield different Ca^{2+} signals at the PF synapses. For example, there are indirect indications of supralinear postsynaptic signal summation during coactivation of CFs and PFs (Batchelor and Garthwaite 1997).

Although it is clear that PF activity capable of producing LTD is accompanied by a rise in dendritic $[Ca^{2+}]$, the site of the coincidence detector for signal integration remains to be determined. Although we did not measure the concentration of Ca^{2+} in individual dendritic spines in the present experiments, in an earlier study we demonstrated that PF activity increases Ca^{2+} levels in both the dendritic shaft and spines (Eilers et al. 1995a). It is therefore likely that induction of LTD is accompanied by sustained increases in Ca^{2+} concentration in both dendritic shafts and spines. Further work will be needed to determine whether the biochemical machinery involved in LTD induction requires the buildup of Ca^{2+} in the dendritic shaft, spines, or both. A related open question is whether there is a threshold in $[Ca^{2+}]_i$ for induction of LTD or whether graded synaptic activity, and the associated graded Ca^{2+} signals (Eilers et al. 1995a), produce proportional reductions in the efficacy of synaptic transmission of PFs.

Various Ca^{2+} sources may contribute to the signals occurring during the induction of PF-LTD. Both AMPA-type glutamate receptor channels (AMPA-Rs) and metabotropic glutamate receptors (mGlu-Rs) are activated during PF synaptic transmission (Konnerth et al. 1990; Batchelor et al. 1994). Ca^{2+} influx through postsynaptic AMPA-Rs has been implicated in a form of synaptic plasticity in dorsal root neurons (Gu et al. 1996). However, because Purkinje cell AMPA-Rs have a very low permeability to Ca^{2+} (Tempia et al. 1996), it is unlikely that Ca^{2+} influx through these receptors plays a significant role in PF-LTD. It is more likely that AMPA-R activation promotes Ca^{2+} influx by activating dendritic voltage-gated Ca^{2+} channels (VGCCs). Both T-type (Mouginot et al. 1997) and P-type (Regan 1991; Usowicz et al. 1992) VGCCs are present in Purkinje cell dendrites; the precise roles of these Ca^{2+} channel subtypes in PF-mediated Ca^{2+} signaling and in the induction of PF-LTD remains to be determined. Activation of type-I mGlu-Rs, the subtype located at PF–Purkinje cell synapses, produces IP_3. Thus, Ca^{2+} release by IP_3 (Llano et al. 1991b; E.A. Finch and G.J. Augustine, unpubl.), as well as Ca^{2+}-induced Ca^{2+} release (Kano et al. 1995), may also contribute to the local dendritic Ca^{2+} signals that induce PF-LTD.

The coincidence detection mechanism for CF and depolarization-mediated LTD requires backpropagation of electrical activity into dendrites. In contrast, we have found that PF-LTD can be induced in the absence of somatic regenerative electrical activity. PF-LTD is instead a purely dendritic form of integration, based on the coincidence of PF activity and a concomitant local increase in dendritic $[Ca^{2+}]_i$. PF-LTD thus represents a striking example of local dendritic signal processing in the mammalian brain.

Acknowledgments

This work was supported by Deutsche Forschungsgemeinschaft grant SFB 246 to A.K. and J.E. and a Bundeministerium für Bildung, Forschung, und Technologie grant to A.K., as well as National Institutes of Health (NIH) Fellowship NS-09586 and a Grass Fellowship to E.A.F. and NIH grant NS-34045 to G.J.A.

The publication costs of this article were defrayed in part by payment of page charges. This article must therefore be hereby marked ''advertisement'' in accordance with 18 USC section 1734 solely to indicate this fact.

References

Aiba, A., M. Kano, C. Chen, M.E. Stanton, G.D. Fox, K. Herrup, T.A. Zwingman, and S. Tonegawa. 1994. Deficient cerebellar long-term depression and impaired motor learning in mGluR1 mutant mice. *Cell* 79: 377–388.

Batchelor, A. and J. Garthwaite. 1997. Frequency detection and temporally dispersed synaptic signal association through a metabotropic glutamate receptor pathway. *Nature* **385:** 74–77.

Batchelor, A.M., D.J. Madge, and J. Garthwaite. 1994. Synaptic activation of metabotropic glutamate receptors in the parallel fibre-Purkinje cell pathway in rat cerebellar slices. *Neuroscience* **63:** 911–915.

Bourne, H.R. and R. Nicoll. 1993. Molecular machines integrate coincident synaptic signals. *Cell* **72:** 65–75.

Crepel, F. and D. Jaillard. 1991. Pairing of pre- and postsynaptic activities in cerebellar Purkinje cells induces long-term changes in synaptic efficacy in vitro. *J. Physiol.* **432:** 123–141.

Crepel, F. and M. Krupa. 1988. Activation of protein kinase C induces a long-term depression of glutamate sensitivity of cerebellar Purkinje cells. An in vitro study. *Brain Res.* **458:** 397–401.

Denk, W., M. Sugimori, and R. Llinás. 1995. Two types of calcium responses limited to single spines in cerebellar Purkinje cells. *Proc. Natl. Acad. Sci.* **92:** 8279–8282.

De Schutter, E. 1995. Cerebellar long-term depression might normalize excitation of Purkinje cells: a hypothesis. *Trends Neurosci.* **18:** 291–295.

De Schutter, E. and R. Maex. 1996. The cerebellum: cortical processing and theory. *Curr. Opin. Neurobiol.* **6:** 759–764.

Edwards, F., A. Konnerth, B. Sakmann, and T. Takahashi. 1989. A thin slice preparation for patch clamp recordings from neurones of the mammalian central nervous system. *Pflügers Arch.* **414:** 600–612.

Eilers, J., G.J. Augustine, and A. Konnerth. 1995a. Subthreshold synaptic Ca^{2+} signalling in fine dendrites and spines of cerebellar Purkinje neurons. *Nature* **373:** 155–158.

Eilers, J., R. Schneggenburger, and A. Konnerth. 1995b. Patch clamp and calcium imaging in brain slices. In *Single channel recording* (ed. B. Sakmann, and E. Neher), pp. 213–229. Plenum Publishing Co., New York, NY.

Gu, J., C. Albuquerque, C. Lee, and A. MacDermott. 1996. Synaptic strengthening through activation of Ca^{2+}-permeable AMPA receptors. *Nature* **381:** 793–796.

Hartell, N.A. 1994. Induction of cerebellar long-term depression requires activation of glutamate metabotropic receptors. *NeuroReport* **5:** 913–916.

———. 1996a. Strong activation of parallel fibers produce localized calcium transients and a form of LTD that spreads to distant synapses. *Neuron* **16:** 601–610.

———. 1996b. Inhibition of cGMP breakdown promotes the induction of cerebellar long-term depression. *J. Neurosci.* **16:** 2881–2890.

Ito, M. 1984. *The cerebellum and neuronal control.* Raven Press, New York, NY.

———. 1989. Long-term depression. *Annu. Rev. Neurosci.* **12:** 85–102.

———. 1991. The cellular basis of cerebellar plasticity. *Curr. Opin. Neurobiol.* **1:** 616–620.

Ito, M. and M. Kano. 1982. Long-lasting depression of parallel fiber-Purkinje cell transmission induced by conjunctive stimulation of parallel fibers and climbing fibers in the cerebellar cortex. *Neurosci. Lett.* **33:** 253–258.

Ito, M., M. Sakurai, and P. Tongroach. 1982. Climbing fibre induced depression of both mossy fibre responsiveness and glutamate sensitivity of cerebellar Purkinje cells. *J. Physiol.* **324:** 113–134.

Kano, M., O. Garaschuk, A. Verkhrtatsky, and A. Konnerth. 1995. Ryanodine receptor-mediated intracellular calcium release in rat cerebellar Purkinje neurones. *J. Physiol.* **487.1:** 1–16.

Kasono, K. and T. Hirano. 1994. Critical role of postsynaptic calcium in cerebellar long-term depression. *NeuroReport* **6:** 17–20.

———. 1995. Involvement of inositol trisphosphate in cerebellar long-term depression. *NeuroReport* **6:** 569–572.

Konnerth, A., I. Llano, and C.M. Armstrong. 1990. Synaptic currents in cerebellar Purkinje cells. *Proc. Natl. Acad. Sci.* **87:** 2662–2665.

Konnerth, A., J. Dreessen, and G.J. Augustine. 1992. Brief dendritic calcium signals initiate long-lasting synaptic depression in cerebellar Purkinje cells. *Proc. Natl. Acad. Sci.* **89:** 7051–7055.

Konnerth, A., R.Y. Tsien, K. Mikoshiba, and J. Altman. 1996. *Coincidence detection in the nervous system.* Human Frontier Science Program, Strasbourg, France.

Krupa, D.J., J.K. Thompson, and R.F. Thompson. 1993. Localization of a memory trace in the mammalian brain. *Science* **260:** 989–991.

Lev-Ram, V., L. Makings, P. Keitz, J. Kao, and R. Tsien. 1995. Long-term depression in cerebellar Purkinje neurons results from coincidence of nitric oxide and depolarization-induced Ca^{2+} transients. *Neuron* **15:** 407–415.

Linden, D.J. 1994. Long-term synaptic depression in the mammalian brain. *Neuron* **7:** 457–472.

Linden, D.J. and J.A. Connor. 1991. Participation of postsynaptic PKC in cerebellar long-term depression in culture. *Science* **254:** 1656–1659.

Linden, D.J., M.H. Dickinson, M. Smeyne, and J.A. Connor. 1991. A long-term depression of AMPA currents in cultured cerebellar Purkinje neurons. *Neuron* **7:** 81–89.

Llano, I., A. Marty, C.M. Armstrong, and A. Konnerth. 1991a. Synaptic- and agonist-induced excitatory currents of Purkinje cells in rat cerebellar slices. *J. Physiol.* **434:** 183–213.

Llano, I., J. Dreessen, M. Kano, and A. Konnerth. 1991b. Intradendritic release of calcium induced by glutamate in cerebellar Purkinje cells. *Neuron* **7:** 577–583.

Magee, J.C., G. Christofi, H. Miyakawa, B. Christie, N. Lasser-Ross, and D. Johnston. 1996. Subthreshold synaptic activation of voltage-gated Ca^{2+} channels mediates a localized Ca^{2+} influx into the dendrites of hippocampal pyramidal neurons. *J. Neurophysiol.* **74:** 1335–1342.

Miyakawa, H., V. Lev-Ram, N. Lasser-Ross, and W.N. Ross. 1992. Calcium transients evoked by climbing fiber and parallel fiber synaptic inputs in Guinea pig cerebellar Purkinje neurons. *J. Neurophysiol.* **68:** 1178–1189.

Mouginot, D., J.-L. Bossu, and B.H. Gähwiler. 1997. Low-threshold Ca^{2+} currents in dendritic recordings from Purkinje cells in rat cerebellar slice cultures. *J. Neurosci.* **17:** 160–170.

Neher, E and G.J. Augustine. 1992. Calcium gradients and buffers in bovine chromaffin cells. *J. Physiol.* **450:** 273–301.

Pusch, M. and E. Neher. 1988. Rates of diffusional exchange between small cells and a measuring patch pipette. *Pflügers Arch.* **411:** 204–211.

Regan, L.J. 1991. Voltage-dependent calcium currents in Purkinje cells from rat cerebellar vermis. *J. Neurosci.* **11:** 2259–2269.

Rexhausen, U. 1992. "Bestimmung der Diffusionseigenschaften von Fluoreszenzfarbstoffen in verzweigten Nervenzellen unter Verwendung eines rechnergesteuerten Bildverarbeitungssystems." Ph.D. Thesis, University of Göttingen, Germany.

Ross, W.N. and R. Werman. 1987. Mapping calcium transients in the dendrites of Purkinje cells from the Guinea-pig cerebellum in vitro. *J. Physiol.* **389:** 319–336.

Sakurai, M. 1990. Calcium is an intracellular mediator of the climbing fiber in induction of cerebellar long-term depression. *Proc. Natl. Acad. Sci.* **87:** 3383–3385.

Tempia, F., M. Kano, R. Schneggenburger, C. Schirra, O. Garaschuk, T. Plant, and A. Konnerth. 1996. Fractional calcium current through neuronal AMPA-receptor channels with a low calcium permeability. *J. Neurosci.* **16:** 456–466.

Usowicz, M.M., M. Sugimori, B. Cherksey, and R. Llinás. 1992. P-type calcium channels in the somata and dendrites of adult cerebellar Purkinje cells. *Neuron* **9:** 1185–1199.

Received March 17, 1997; accepted in revised form May 2, 1997.

Absence Of Cerebellar Long-Term Depression in Mice Lacking Neuronal Nitric Oxide Synthase

Varda Lev-Ram,[1] Zuryash Nebyelul,[2] Mark H. Ellisman,[2] Paul L. Huang,[3] and Roger Y. Tsien[1,4,5]

[1]Department of Pharmacology
[2]Department of Neurosciences
and [4]Howard Hughes Medical Institute
University of California San Diego
La Jolla, California 92093-0647
[3]Cardiovascular Research Center
Massachusetts General Hospital
Charlestown, Massachusetts 02129-2060

Abstract

Extensive pharmacological evidence suggests that nitric oxide (NO) is a crucial transmitter for cerebellar long-term depression (LTD), a long-lasting decrease in efficacy of the synapses from parallel fibers onto Purkinje neurons, triggered by coincident presynaptic activity and postsynaptic depolarization. We now show that LTD cannot be induced in Purkinje neurons under whole-cell patch clamp in cerebellar slices from young adult mice genetically lacking neuronal nitric oxide synthase (nNOS). This genetic evidence confirms the essentiality of NO and nNOS for LTD in young adult rodents. Surprisingly, LTD in cells from nNOS knockout mice cannot be rescued by photolytic uncaging of NO and cGMP inside Purkinje neurons, although such stimuli circumvent acute pharmacological inhibition of nNOS and soluble guanylate cyclase in normal rodents. Also slices from knockout mice show no deficit in cGMP elevation in response to exogenous NO. Therefore, prolonged absence of nNOS allows atrophy of the signaling pathway downstream of cGMP.

[5]Corresponding author.

Introduction

The gaseous messenger nitric oxide (NO) has recently been recognized as a modulatory neurotransmitter in the central nervous system (Madison 1993; Bredt and Snyder 1994). NO is made from arginine by the enzyme nitric oxide synthase (NOS). At least three isoforms of the enzyme are known, neuronal (nNOS), endothelial (eNOS) and inducible (iNOS). nNOS is particularly abundant in the cerebellum (Bredt et al. 1990), where NO is crucial for the induction of long-term depression (LTD), a reduction in efficacy of the synapses from parallel fibers (PFs) onto Purkinje cells (PCs). LTD is physiologically produced by simultaneous stimulation of PF and climbing fiber (CF) inputs to the PC. A rise in free Ca^{2+} concentration ($[Ca^{2+}]_i$) inside the PC is both sufficient and necessary to mediate the effect of the CF, because CF stimulation can be entirely replaced either by simple depolarization (Konnerth et al. 1992), which activates voltage-operated Ca^{2+} channels, or by photolytic uncaging of Ca^{2+} from a light-sensitive chelator (Lev-Ram et al. 1997). If $[Ca^{2+}]_i$ is buffered by chelators, then LTD cannot be elicited (Konnerth et al. 1992; Lev-Ram et al. 1995). Meanwhile NO production is both sufficient and necessary to mediate the effect of the PF, because photorelease of caged NO inside the PC completely replaces PF activity and synergizes with either depolarization or uncaged $[Ca^{2+}]_i$ to cause LTD (Lev-Ram et al. 1995, 1997). Inhibition of presynaptic NOS with L-nitroarginine, or trapping of NO by myoglobin either outside or inside the PC, prevents LTD induction through PF activity, whereas introduction of L-nitroarginine directly into PCs had no effect. NO uncaged inside the PC circumvents NOS inhibition or extracellular myoglobin but not intracellular myoglobin, confirming that PF activity must generate NO outside the PC and that NO then has to diffuse into the PC to have its effect (Lev-Ram et al. 1995). The most likely molecular target of NO is soluble guanylate cyclase (sGC), which is very abundant in PCs and is the enzyme that synthesizes cGMP. Inhibition of sGC prevents PF activity or uncaged NO from participating in LTD induction, but such inhibition can be circumvented by uncaged cGMP, which synergizes with depolarization-induced $[Ca^{2+}]_i$ transients to elicit LTD. Therefore, cGMP is downstream of NO and is likewise necessary and sufficient for mediating the PF signal (Lev-Ram et al. 1997).

The previous dissection of the signals for LTD induction was accomplished by pharmacological means, including enzyme inhibitors, Ca^{2+} and NO traps, and caged compounds that release Ca^{2+}, NO, or cGMP upon flash photolysis. Specificity of pharmacological inhibition of NOS or sGC could be confirmed by rescuing the block by uncaging the downstream messenger NO or cGMP. A complementary approach to NO function would be genetic deletion of NOS or sGC. Although sGC knockouts have not yet been reported, mice lacking nNOS have been generated by homologous recombination (Huang et al. 1993), and the homozygotes are phenotypically healthy, but the males have been reported to be abnormally aggressive (Nelson et al. 1995). Likewise eNOS knockouts were also generated (Huang et al. 1995). Neither mouse line showed deficiencies in hippocampal long-term potentiation (LTP) or spatial learning; forms of plasticity that have also been proposed to require NO. However, deletion of both nNOS and eNOS does seem to prevent hippocampal LTP at least in apical dendrites, so the two isoforms seem to substitute for each other in the hippocampus (Son et al. 1996).

The only previous test of nNOS deletion on cerebellar neurons was by Linden et al. (1995), who cultured PCs from embryonic knockout mice and found no deficit in their ability to desensitize in response to simultaneous glutamate application and depolarization. This result confirmed their extensive previous data (including the lack of effect of NO donors or NOS inhibitors) showing that NO is not essential for LTD in cultured PCs (Linden and Connor 1992). However, the pharmacological tests already discussed show that LTD in cerebellar slices freshly prepared from young adult rodents is quite different from the desensitization in PCs isolated from much younger animals and maintained in culture for prolonged periods. Therefore, it was important to compare brain slices from young adult mice with and without nNOS for their ability to show LTD. If NO is as important as our previous results indicated, and if nNOS is the crucial isozyme responsible for NO production, then genetic deletion of nNOS should prevent LTD. We now show this prediction to be correct. Surprisingly, neither photoreleased NO nor uncaged cGMP, both combined with depolarization, could rescue LTD induction. The most likely explanation is that the absence of nNOS throughout development has allowed signaling components downstream of cGMP to atrophy.

Materials and Methods

The homozygous mice with targeted deletion of their nNOS gene have been described previously (Huang et al. 1993). Their appearance and behavior were at least superficially quite normal, and the males showed no sign of the previously reported aggressiveness (Nelson et al. 1995). Therefore, the absence of nNOS in the cerebellum was verified by immunocytochemistry (see below and Fig. 1).

ELECTROPHYSIOLOGY

Thin (300 µm) sagittal slices were cut with a Microslicer DSK-3000W (Dosaka EM Co., Japan) from the cerebellar vermis of mice aged 18–21 days. Synaptic currents in PCs were recorded in the whole-cell patch-clamp configuration (Hamill et al. 1981; Edwards et al. 1989). The cells were directly visualized through a 10× water immersion objective on an upright microscope (Axioplan, Zeiss). Tight-seal whole-cell recordings (seal resistance > 10 GΩ) were made with patch pipettes with 3–4 MΔ resistance and an Axopatch 200A (Axon Instruments) amplifier (holding potential, −70 mV). A 10-msec, 10-mV test depolarization preceded each PF stimulus to monitor the input resistance of the PC throughout the experiment, which was discarded if the resistance changed by >20%. The intracellular solution contained mM K-gluconate; 10 mM KCl; 10 mM K-HEPES; 1 mM

MgCl$_2$; 4 mM Na-ATP; 1 mM Na-GTP; 16 mM sucrose; (pH 7.2) osmolarity 300 Osm. Although these conditions are adequate for measuring excitatory postsynaptic currents (EPSCs), the voltage clamp is incapable of maintaining space clamp conditions in the dendrites and axon when spikes are generated. The external Ringer's solution contained 125 mM NaCl; 2.5 mM KCl; 2 mM CaCl$_2$; 1 mM MgCl$_2$; 1.25 mM NaH$_2$PO$_4$; 26 mM NaHCO$_3$; 25 mM glucose (pH 7.4); and 10 μM (−)-bicuculline methiodide (Research Biochemical International) to inhibit GABAergic synapses. Slices were continually perfused with Ringer's solution saturated with 95% O$_2$, 5% CO$_2$, during recording. All experiments were performed at room temperature (near 22°C).

Intracellular calcium increases were achieved by 50-msec step depolarizations to a level in which regenerative calcium spikes were induced in the out-of-clamp dendritic region. For PF stimulation a bipolar electrode was placed at the pial surface above and up to 100 μm on either side of the recorded PC. A 50-μsec pulse was delivered to stimulate the PF using an isolation unit (ISO-Flex, A.M.P.I., Israel). The timing of depolarization, PF stimulation, and shutter opening was orchestrated using a multichannel stimulator (Master-8, A.M.P.I., Israel). Voltage and current data were digitized and stored on a VCR recorder via a PCM2 A/D VCR adaptor (Medical System Corp., Greenvale, NY) and transferred via a PCMI2 interface to a personal computer for software extraction of synaptic current amplitudes.

ODQ (1H-[1,2,4]oxadiazolo[4,3-a]quinoxalin-1-one) was purchased from Tocris Cookson Inc. (Sigma, St. Louis, MO). The spontaneous and caged NO donors Et$_2$N(N$_2$O$_2$)Na, CNO-4 dipotassium salt (Makings and Tsien 1994), and the caged cGMP CM–cGMP [guanosine 3′,5′-cyclic monophosphate, 4,5-bis(carboxymethoxy)-2-nitrobenzyl ester] (Lev-Ram et al. 1997) were synthesized in this laboratory. Caged compounds were introduced into the Purkinje neuron by inclusion in the patch pipette intracellular solution. For further details of the uncaging apparatus, see Lev-Ram et al. (1995, 1997).

cGMP IMMUNOASSAY

cGMP assays were performed on 300-μm-thick slices, which were incubated in oxygenated Ringer's solution throughout the experiment. The slices were preincubated with control Ringer, ODQ (8 μM), or IBMX (20 μM) for 10 min before stimulation with control Ringer, Et$_2$N(N$_2$O$_2$)Na (100 μM), or glutamate (2 μM) for 3 min. At the end of the stimulation period the slices were dropped into liquid nitrogen. After all the slices were frozen, 1 ml of 5% TCA was added and the slices were sonicated for 5 min. Each slice was triturated and then transferred to a microcentrifuge tube and centrifuged in the cold for 5 min. The supernatant was collected and assayed for cGMP using an enzyme immunoassay kit (Cayman Chemical) following the procedure described in the kit instruction booklet. Each sample consisted of one or two slices, whose protein content was determined by the bicinchoninic acid method. The values of picomoles of cGMP/mg of protein were averaged for the two duplicate tissue samples that had been subjected to each set of preincubation and stimulus conditions, then normalized by the picomoles of cGMP/mg of protein for the samples with no inhibitors or stimulators. The resulting cGMP amplification factors were finally averaged for three separate experiments; the bars indicate the standard errors of the final averaging (Fig. 4).

IMMUNOHISTOCHEMISTRY

Mice were anesthetized with 4 ml/kg of body weight ketamine/Rompun and perfused using intracardiac catheterization. Perfusion with a balanced salt solution (135 mM NaCl, 14 mM NaHCO$_3$, 1.2 mM Na$_2$HPO$_4$, 5 mM KCl, 2 mM CaCl$_2$, 1 mM MgCl$_2$) at 35°C was followed by 4% formaldehyde (fresh from paraformaldehyde) in 0.1 M phosphate buffered saline (PBS) at pH 7.4 for 5 min. The cerebelli were removed and fixed an additional 1 hr, then sections 40–60 μm thick were cut with a vibratome (Lancer). The sections were permeabilized in a solution containing 0.1% Triton X-100, 1% normal goat serum, and 1% cold water fish gelatin (Sigma) in PBS for 30 min before incubation in affinity-purified anti-NOS antibody (Transductions Laboratory) at a dilution of 1:50 for 18 hr at 4°C. Sections were then washed in buffer and incubated in goat anti-rabbit IgG–FITC conjugate (Jackson ImmunoResearch) for 1 hr at 4°C. Following this, the sections were rinsed in PBS and mounted in Gelvatol. Confocal microscopy was performed using an MRC-1024 confocal microscope system (Bio-Rad) attached to an Axiovert 35M (Zeiss) using a 40× 1.3 N.A. objective. Excitation was 488 nm from an argon ion laser. Digital images were printed using a Fujix Pictrography 3000 printer (Fuji).

Results

LOCALIZATION OF nNOS IMMUNOREACTIVITY IN CEREBELLAR SLICES FROM MUTANT AND CONTROL MICE

To confirm that the mutant mice are lacking nNOS, immunohistochemical localization of the enzyme was performed on cerebellar slices from knockout mice and C57Bl/6J as normal controls. nNOS immunoreactivity was present in the control C57Bl/6J cerebellum in non-PCs (Fig. 1A,B; see also Bredt et al. 1990) but was missing in the mutant as expected (Fig. 1C).

LTD INDCUTION BY PF STIMULATION AND DEPOLARIZATION

LTD in acute cerebellar slices can be induced by stimulation of PFs synchronized with a 50-msec postsynaptic depolarization repeated at 1 Hz for 30 sec. Figure 2A demonstrates the effectiveness of this paradigm in causing LTD in six PCs from control mice. However, the same protocol (a) was completely ineffective in eight cells from nNOS knockout mice (Fig. 2C).

LTD INDUCTION BY PHOTORELEASED NO AND DEPOLARIZATION

If the only effect of genetic deletion of nNOS were the inability to generate NO, uncaged NO should bypass this deficit just as it does for L-nitroarginine block of NOS. Therefore, caged NO (CNO-4) (Makings and Tsien 1994; Lev-Ram et al. 1995) was included in the patch pipette (200 μM) and photolyzed in concert with a depolarization-induced calcium transient at 1 Hz for 30 sec. This protocol (b) by itself produced LTD in normal rats (Lev-Ram et al. 1995) and mice (Fig. 2B) but failed to induce LTD in any of the eight PCs from mutant mice (Fig. 2C). The inability to produce LTD was not confined to the particular dose of uncaged NO used in Figure 2C, because longer and shorter UV exposures to produce larger and smaller amounts of photoreleased NO were still ineffective in the mutant cells (data not shown). Thus, the lack of nNOS during development had presumably inter-

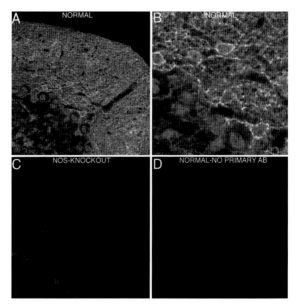

Figure 1: Localization in nNOS immunoreactivity in the cerebellum of normal and nNOS-deficient mice. nNOS immunoreactivity in normal (C57BL/6J) mice cerebellar vermis indicates the expected distribution in non-PCs (A). (B) A higher magnification of an area in A. (C) The lack of nNOS in the mutant mice. (D) No fluorescent labeling can be detected in the absence of the primary antibody. (A,C,D) 164 μm across; (B) 80 μm across.

fered with the expression, localization, or function of downstream elements in the signal transduction pathway.

LTD INDUCTION BY PHOTORELEASED cGMP AND DEPOLARIZATION

To assess whether the additional deficit in nNOS-deficient mice is upstream or downstream of cGMP, we asked whether LTD can be induced by photoreleased cGMP combined with depolarization. In slices from normal rats (Lev-Ram et al. 1997) and mice (Fig. 3B), this protocol reliably induced LTD. The same protocol failed to give LTD in nNOS-deficient mice (Fig. 3C). In five out of eight PCs, the EPSC amplitudes actually increased somewhat during and after the stimulation protocol, much like the behavior of normal rat PCs in which cGMP-dependent protein kinase was pharmacologically inhibited (Lev-Ram et al. 1997). Even in the three nNOS-deficient PCs in which the baseline remained flat, for example, the cell shown in Figure 3D, there was no sign of LTD. Therefore, the PCs from the mutant mice show a deficit in sensing cGMP, in addition to their lack of nNOS.

cerebellar long-term depression and impaired motor learning in mGluR1 mutant mice. *Cell* **79**: 377–388.

Bacskai, B.J., B. Hochner, M. Mahaut-Smith, S.R. Adams, B.-K. Kaang, E.R. Kandel, and R.Y. Tsien. 1993. Spatially resolved dynamics of cAMP and protein kinase A subunits in *Aplysia* sensory neurons. *Science* **260**: 222–226.

Bredt, D.S. and S.H. Snyder. 1994. Nitric oxide: A physiologic messenger molecule. *Annu. Rev. Biochem.* **63**: 175–195.

Bredt, D.S., P.M. Hwang, and S.H. Snyder. 1990. Localization of nitric oxide synthase indicating a neural role for nitric oxide. *Nature* **347**: 768–770.

Crosby, G., J.J.A. Marota, and P.L. Huang. 1995. Intact nociception-induced neuroplasticity in transgenic mice deficient in neuronal nitric oxide synthase. *Neuroscience* **69**: 1013–1017.

Dawson, V.L., V.M. Kizushi, P.L. Huang, S.H. Snyder, and T.M. Dawson. 1996. Resistance to neurotoxicity in cortical cultures from neuronal nitric oxide synthase-deficient mice. *J. Neurosci.* **16**: 2479–2487.

Edwards, F.A., A. Konnerth, B. Sakmann, and T. Takahashi. 1989. A thin slice preparation for patch clamp recordings from neurons of the mammalian central nervous system. *Pflügers Arch.* **414**: 600–612.

Hamill, O.P., A. Marty, E. Neher, B. Sakmann, and F.J. Sigworth. 1981. Improved patch-clamp techniques for high-resolution current recording from cells and cell-free membrane patches. *Pflügers Arch.* **391**: 85–100.

Hempel, C.M., P. Vincent, S.R. Adams, R.Y. Tsien, and A.I. Selverston. 1996. Spatio-temporal dynamics of cAMP signals in an intact neural circuit. *Nature* **384**: 166–169.

Huang, P.L., T.M. Dawson, D.S. Bredt, S.H. Snyder, and M.C. Fishman. 1993. Targeted disruption of the neuronal nitric oxide synthase gene. *Cell* **75**: 1273–1286.

Huang, P.L., Z. Huang, H. Mashimo, K.D. Bloch, M.A. Moskowitz, J.A. Bevan, and M.C. Fishman. 1995. Hypertension in mice lacking the gene for endothelial nitric oxide synthase. *Nature* **377**: 239–242.

Irikura, K., P.L. Huang, J. Ma, W.S. Lee, T. Dalkara, M.C. Fishman, T.M. Dawson, S.H. Snyder, and M.A. Moskowitz. 1995. Cerebrovascular alterations in mice lacking neuronal nitric oxide synthase gene expression. *Proc. Natl. Acad. Sci.* **92**: 6823–6827.

Konnerth, A., J. Dreessen, and G.J. Augustine. 1992. Brief dendritic calcium signals initiate long-lasting synaptic depression in cerebellar Purkinje cells. *Proc. Natl. Acad. Sci.* **89**: 7051–7055.

Lev-Ram, V., L.R. Makings, P.F. Keitz, J.P.Y. Kao, and R.Y. Tsien. 1995. Long-term depression in cerebellar Purkinje neurons results from coincidence of nitric oxide and depolarization-induced Ca^{2+} transients. *Neuron* **15**: 407–415.

Lev-Ram, V., T. Jiang, J. Wood, D.S. Lawrence, and R.Y. Tsien. 1997. Synergies and coincidence requirements between NO, cGMP, and Ca^{2+} in the induction of cerebellar long-term depression. *Neuron* (in press).

Linden, D.J. and J.A. Connor. 1992. Long-term depression of glutamate currents in cultured cerebellar Purkinje neurons does not require nitric oxide signaling. *Eur. J. Neurosci.* **4**: 10–15.

Linden, D.J., T.M. Dawson, and V.L. Dawson. 1995. An evaluation of the nitric oxide/cGMP/cGMP-dependent protein kinase cascade in the induction of cerebellar long-term depression in culture. *J. Neurosci.* **15**: 5098–5105.

Madison, D.V. 1993. Pass the nitric oxide. *Proc. Natl. Acad. Sci.* **90**: 4329–4331.

Makings, L.R. and R.Y. Tsien. 1994. Caged nitric oxide: Stable, organic molecules from which nitric oxide can be photoreleased. *J. Biol. Chem.* **269**: 6282–6285.

Maragosa, C.M., D. Morley, D.A. Wink, T.M. Dunams, J.E. Saavedra, A. Hoffman, A.A. Bove, L. Issac, J.A. Hrabie, and L.K. Keefer. 1991. Complexes of NO with nucleophiles as agents for the controlled biological release of nitric oxide. Vasorelaxant effects. *J. Med. Chem.* **34**: 3242–3247.

Nelson, R.J., G.E. Demas, P.L. Huang, M.C. Fishman, V.L. Dawson, T.M. Dawson, and S.H. Snyder. 1995. Behavioural abnormalities in male mice lacking neuronal nitric oxide synthase. *Nature* **378**: 383–386.

O'Dell, T.J., P.L. Huang, T.M. Dawson, J.L. Dinerman, S.H. Snyder, E.R. Kandel, and M.C. Fishman. 1994. Endothelial NOS and the blockade of LTP by NOS inhibitors in mice lacking neuronal NOS. *Science* **265**: 542–546.

Shibuki, K., H. Gomi, L. Chen, S. Bao, J.J. Kim, H. Wakatsuki, T. Fujisaki, K. Fujimoto, A. Katoh, T. Ikeda, C. Chen, R.F. Thompson, and S. Itohara. 1996. Deficient cerebellar long-term depression, impaired eyeblink conditioning, and normal motor coordination in GFAP mutant mice. *Neuron* **16**: 587–599.

Son, H., R.D. Hawkins, K. Martin, M. Kiebler, P.L. Huang, M.C. Fishman, and E.R. Kandel. 1996. Long-term potentiation is reduced in mice that are doubly mutant in endothelial and neuronal nitric oxide synthase. *Cell* **87**: 1015–1023.

Received May 19, 1997; accepted in revised form May 28, 1997.

INFORMATION FOR CONTRIBUTORS (1997)

Aims and Scope

LEARNING & MEMORY welcomes high-quality original research papers on all types of learning, memory, and their models, conducted in humans and in vertebrate and invertebrate species with the following approaches: behavior, cognition, neuroanatomy, neurophysiology, neuropharmacology, biochemistry, genetics, and cell and molecular biology. The journal will also publish review articles, theoretical papers, and short communications, including comments on published papers and the authors' responses.

Submission of Papers

The journal accepts papers that present original unpublished research. Submission to the journal implies that a paper is not currently being considered for another journal or book. Closely related papers that are in press elsewhere or that have been submitted elsewhere should be included with the submitted manuscript. It is also understood that researchers who submit papers to this journal are prepared to make available to qualified academic researchers materials needed to duplicate their research results. Authors should submit nucleic acid and protein sequences to the appropriate data bank. Questions regarding papers should be directed to Judy Cuddihy, Managing Editor, at Cold Spring Harbor Laboratory (516-367-8492).

Manuscripts should be submitted to:
Judy Cuddihy, Managing Editor
Learning & Memory
Cold Spring Harbor Laboratory
1 Bungtown Road
Cold Spring Harbor, New York 11724

Manuscript Preparation

1. General. Papers should be as concise as possible. The entire paper (including tables, figure legends, references, footnotes) should be typed double-spaced on standard-sized European or American bond paper with at least 1-in (2.5 cm) margins on all four sides. Computer printouts should be of letter quality, and should use a computer typeface of at least 11 point size. Each page should be labeled with the first author's name and a page number. Five copies should be submitted; at least four of these copies should have original art. A cover letter should include: (a) name, address, and telephone and FAX numbers of author responsible for correspondence regarding the manuscript; (b) statement that the manuscript has been seen and approved by all listed authors; (c) specific requirements for reproduction of art; and (d) status of any permissions needed. Five copies of the manuscript should be submitted for use by referees and editors.

2. Submitted Papers on Computer Discs. Publication can be speeded up if accepted papers are supplied on 3 1/2- or 5 1/4-inch floppy discs. We can accept IBM PC, MacIntosh, or compatible, formats. Please supply the manuscript on the disc as a "text" or ASCII file, if possible. Please indicate on the disc: computer brand name, whether the disc contains a text or word-processing file (name of software), and the disc format. Five hard copy versions should also be submitted for use by referees and editors.

3. Forms. The following order is preferred: Title page, Abstract, Introduction, Methods, Results, Discussion, Acknowledgments, References, Tables, Figure legends. The Title page should include : (a) title; (b) all authors' full names; (c) all affiliations clearly indicated; (d) a shortened version of the title for use as a running head (maximum 45 characters); and (e) key words (up to 6) for use in indexing. The Abstract should be about 200 words long and should summarize the aim of the report, the methodological approach, and the significance of the results. Methods should be detailed enough to allow any qualified researcher to duplicate the results.

4. Figures and Legends. Five sets of figures should be supplied as high-quality glossy prints. Half-tones should be high-contrast, particularly in the case of gels, for the best reproduction. Line drawings, graphs, charts, and chemical formulae, should be professionally prepared and labeled. Multiple-part figures should be submitted as mounted, camera-ready composites. Authors submitting color figures as essential data for review with manuscripts undertake to pay the publication costs of four-color artwork. Price estimates are supplied upon acceptance of the paper.

The back of each figure should be labeled with the first author's name, figure number, and an indication of "top." The figures should be numbered consecutively in the order to which they are referred in the text. The size of the figures will be adjusted to fit the Journal format, therefore please try to keep labels, symbols, and other call-out devices in proportion to the figure size detail.

Figure legends should be brief and should not contain methods. Symbols indicated in the figure should be identified in the legend text. If figures are reprinted from another source, permission to reprint is required.

5. Tables. Tabular data should be presented concisely and logically. Tables should be numbered consecutively according to the order cited in the text and each should have a title. Use only horizontal rules and make sure column headings are unambiguous in indicating columns to which they refer. Table legends and footnotes should be included where needed. If tables are reprinted from another source or if data included are from another source, permission to reprint is required.

6. References. References are name/date citations in text; please do not cite by number. Undated citations (unpublished, in preparation, personal communication) should include first initials and last names of authors. Bibliographic information should be supplied in the following order: For Journal articles; Author(s), year, article title, journal title, volume, inclusive page numbers. For books: Author(s), year, chapter title, book title, editors' names, volume, inclusive page numbers, publisher, city of publication.

7. Proofs. Proofs are considered the final form of the paper and correction can be made only in the case of factual errors. If additional information must be added at this stage, it should be in the form of "Note added in proof," subject to the approval of the editors.

8. Reprints. A reprint order form will be included with the proofs.

To help defray the cost of publication, a charge of $25 per page will be made for publication in Learning & Memory. Authors unable to meet these charges should include a letter of explanation upon acceptance for publication; inability to meet these charges will have no effect on acceptance and publication of submitted papers.